GREAT BY CHOICE

ALSO BY JIM COLLINS

Beyond Entrepreneurship (with William C. Lazier)
Managing the Small to Mid-Sized Company (with William C. Lazier)
Built to Last (with Jerry I. Porras)
Good to Great
Good to Great and the Social Sectors
How the Mighty Fall

ALSO BY MORTEN T. HANSEN

Collaboration

GREAT BY CHOICE

UNCERTAINTY, CHAOS, AND LUCK—WHY SOME THRIVE DESPITE THEM ALL

Jim Collins
AND
Morten T. Hansen

RANDOM HOUSE BUSINESS

007

Random House Business
20 Vauxhall Bridge Road
London SW1V 2SA

Random House Business is part of the Penguin Random House group of companies whose addresses can be found at global.penguinrandomhouse.com

Penguin
Random House
UK

First published in the USA in 2011 by Harper Business, an imprint of HarperCollins Publishers, New York.
First published in the United Kingdom in 2011 by Random House Business Books.

www.penguin.co.uk

A CIP catalogue record for this book is available from the British Library.

ISBN 9781847940889

Printed and bound in India by Replika Press Pvt. Ltd.

Penguin Random House is committed to a sustainable future for our business, our readers and our planet. This book is made from Forest Stewardship Council® certified paper.

FSC
www.fsc.org
MIX
Paper | Supporting responsible forestry
FSC® C018179

FROM JIM:

To my grandmother Delores,
who at age 97 still had big dreams and audacious goals.

FROM MORTEN:

To my daughters, Alexandra and Julia,
whose generation will create the future.

CONTENTS

ACKNOWLEDGMENTS

We could not have completed this project without the small army of people who made significant contributions of time and intellect.

We had a wonderful team of research assistants. They are smart, curious, irreverent, fanatically disciplined people who are a joy to work with. We would like to thank the following members of the Chimp-Works research team: Robyn Bitner for multiple years of marching on a range of analyses, Kyle Blackmer for turbulence insights, Brad Caldwell for Biomet and Southwest analyses, Adam Cederberg for company selection and IPO analyses, Lauren Cujé for 10X-company updates and title analysis, Terrence Cummings (a.k.a. Grande) for thousands of hours invested in dozens of pieces of the project, Daniel DeWispelare for Amgen analysis, Todd Driver for 10X-leader analyses and 10X-company updates, Michael Graham for comparison selection and comparative analyses, Eric Hagen for the IPO-list SMaC check and his big-brain contributions, Ryan Hall for a range of quantitative analyses, Beth Hartman for turbulence analyses and company selections, Deborah Knox for industry-turbulence analyses and extensive IPO analysis, Betina Koski for industry-turbulence analyses, Michael Lane for comparison selection and comparative analyses, Lorilee Linfield for company updates and multiple years of SMaC work, Nicholas M. Osgood for industry-turbulence analyses, Catherine Patterson for comparison selections and comparative analyses, Matthew Unangst for backup analyses and Moore's Law research, and Nathaniel (Natty) Zola for becoming the guru of Southwest Airlines versus Pacific Southwest Airlines (PSA). From Morten's research assistants, we thank Chris

Allen for data analyses, Muhammad Rashid Ansari for industry analyses, Jayne Brocklehurst for research support, Attrace Yuiying Chang for fact checking, Hendrika Escoffier for research support, Roisin Kelly for research support, Chittima Silberzahn for financial data and analyses, Philippe Silberzahn for Microsoft and Apple analyses, William Simpson for data analyses, Gina Carioggia Szigety for data analyses in company selections, Nana von Bernuth for years of amazing effort and commitment in conducting a wide range of indispensable analyses, and James Zeitler for data analyses.

We are indebted to our critical readers who invested hours in reading drafts of the manuscript, criticizing the work, offering suggestions, and pushing at every turn to make the work better. For their candor, insight, and perspective, we would like to thank Ron Adner, Joel T. Allison, FACHE, Chris Barbary, Gerald (Jerry) Belle, Darrell Billington, Kyle Blackmer, John M. Bremen, William P. Buchanan, Scott Calder, Robin Capehart, Scott Cederberg, Brian Cornell, Lauren Cujé, Jeff Donnelly, Todd Driver, David R. Duncan, Joanne Ernst, Mike Faith, Andrew Feiler, Claudio Fernández-Aráoz, Andrew Fimiano, Christopher Forman, John Foster, Dick Frost, Itzik Goldberger, Michael Graham, Ed Greenberg, Eric Hagen, Becky Hall, Ryan Hall, Beth Hartman, Liz Heron, John B. Hess, John G. Hill, Kim Hollingsworth Taylor, Thomas F. Hornbein, MD, Lane Hornung, Zane Huffman, Christine Jones, Scott Jones, David D. Kennedy, Alan Khazei, Betina Koski, Eva M. H. Kristensen, Brian C. Larsen, Kyle Lefkoff, Jim Linfield (father of chimp Lorilee), Ed Ludwig, Wistar H. MacLaren, David Maxwell, Kevin McGarvey, MD, MBA, Bill McNabb, Anne-Worley Moelter (SFVG), Michael James Moelter, Clarence Otis Jr., Larry Pensack, Jerry Peterson, Amy Pressman, Sam Presti, Michael Prouting, David P. Rea, Jim Reid, Neville Richardson, Sara Richardson, Kevin Rumon, David G. Salyers, Kim Sanchez Rael, Vijay Sathe, Keegan Scanlon, Dirk Schlimm, William F. Shuster, Anabel Shyers, Alyson Sinclair, Tim Tassopoulos, Kevin Taweel, Jean Taylor, Tom Tierney, Nicole Toomey Davis, Matthew Unangst, Nana von Bernuth, H. Lawrence Webb, David Weekley, Chuck Wexler, Dave Witherow, and Nathaniel (Natty) Zola. We also thank Constance Hale, Jeffrey Martin, and Filipe Simões dos Santos for their special attention to the research-methods section. In addition, we would like to thank Salvatore D. Fazzolari, Denis Godcharles, Ben R. Leedle Jr., Evan Shapiro, Roy M. Spence Jr., and Jim Weddle for helpful dialogue and feedback.

We would also like to thank the Transportation Library at Northwestern University for access to PSA's annual reports; Betty Grebe and Carol Krismann at the University of Colorado William M. White Business Library; the Center for Research in Security Pricing (CRSP), Booth School of Business at the University of Chicago for its quality data and excellent service; Jasjit Singh for patent data and insights; Dennis Bale and Laurie Drawbaugh for the roving office; Leigh Wilbanks for joining our early concept debates; Alex Toll for proofing; Alan Webber for great moments of conversation that sparked key ideas; Jim Logan for enduring the seemingly endless journey; Tommy Caldwell for testing 10Xer ideas on sheer cliff walls; and all of Jim's Personal Band of Brothers. Morten is especially thankful to Harvard Business School, INSEAD, and University of California–Berkeley, where he held academic positions during the time of this research.

We thank Deborah Knox for editing the final text, pushing us for consistency and clarity, continually challenging our ideas, and *zooming out on the big ideas* while also *zooming in on the details*. We thank James J. Robb for his graphics expertise, unbounded creativity, and enduring friendship. We thank Janet Brockett for her creative spark and design genius. We thank Caryn Marooney for being a farseeing guide through treacherous territory. We thank Peter M. Moldave for his dedicated and considered counsel. We thank Hollis Heimbouch for believing in this work early on, tirelessly navigating the changing landscape of publishing, and working in a true spirit of partnership. We thank Peter Ginsberg for his unbroken track record at bringing together creative and unusual arrangements that benefit everyone involved.

We thank members of the ChimpWorks home team, who make it possible for Jim to focus on doing huge creative projects. For their efforts early in the project, we thank Brian J. Bagley, Patrick Blakemore, Taffee

Hightower, Vicki Mosur Osgood, and Laura Schuchat. We thank Jeff Dale for his measured and wise perspective as our strategic paratrooper; Judi Dunckley for her dedication to precision and accuracy (and for her joyful worrying); Joanne Ernst for serving as chairman of the council and for her unrivaled ability to analyze issues and sharpen our thinking; Michael Lane for his dedicated years of being productively irreverent; Sue Barlow Toll for serving as our director of operations and scampering down many a rabbit hole; and Kathy Worland-Turner for being Jim's Right Arm and exercising her wonderful ability to create friends and build relationships. We thank Robyn Bitner and Lorilee Linfield for their heroic dedication to the project in its final year; they brought light and energy to our team, while being OPURs of the first order.

Finally, we owe an incalculable debt to our respective life partners, Joanne Ernst and Hélène Hansen, for their unyielding support, severest criticism, and endurance as we marched through the nine years of this project. This work would not exist without them.

GREAT BY CHOICE

1

THRIVING IN UNCERTAINTY

"We simply do not know what the future holds."

—Peter L. Bernstein[1]

We cannot predict the future. But we can create it.

Think back to 15 years ago, and consider what's happened since, the destabilizing events—in the world, in your country, in the markets, in your work, in your life—that defied all expectations. We can be astonished, confounded, shocked, stunned, delighted, or terrified, but rarely prescient. None of us can predict with certainty the twists and turns our lives will take. Life is uncertain, the future unknown. This is neither good nor bad. It just *is*, like gravity. Yet the task remains: how to master our own fate, even so.

We began the nine-year research project behind this book in 2002, when America awoke from its false sense of stability, safety, and wealth entitlement. The long-running bull market crashed. The government budget surplus flipped back to deficits. The terrorist attacks of September 11, 2001, horrified and enraged people everywhere; and war followed. Meanwhile, throughout the world, technological change and global competition continued their relentless, disruptive march.

All of this led us to a simple question: *Why do some companies thrive in uncertainty, even chaos, and others do not?* When buffeted by tumultuous events, when hit by big, fast-moving forces that we can neither predict nor control, what distinguishes those who perform exceptionally well from those who underperform or worse?

We don't choose study questions. They choose us. Sometimes one of the questions just grabs us around the throat and growls, "I'm not going to release my grip and let you breathe until you answer me!" This study grabbed us because of our own persistent angst and gnawing sense of vulnerability in a world that feels increasingly disordered. The question wasn't just intellectually interesting but personally relevant. And as we spent time with our students and worked with leaders in both the business and social sectors, we sensed the same angst in them. In the intervening years, events have served only to reinforce this sense of unease. What's coming next? All we know is that no one knows.

Yet some companies and leaders navigate this type of world exceptionally well. They don't merely react; they create. They don't merely survive; they prevail. They don't merely succeed; they thrive. They build great enterprises that can endure. We do not believe that chaos, uncertainty, and instability are good; companies, leaders, organizations, and societies do not thrive *on* chaos. But they can thrive *in* chaos.

To get at the question of how, we set out to find companies that started from a position of vulnerability, rose to become great companies with spectacular performance, and did so in unstable environments characterized by big forces, out of their control, fast moving, uncertain, and potentially harmful. We then compared these companies to a control group of companies that failed to become great in the same extreme environments, using the contrast between winners and also-rans to uncover the distinguishing factors that allow some to thrive in uncertainty.

We labeled our high-performing study cases with the moniker "10X" because they didn't merely get by or just become successful. They truly thrived. Every 10X case beat its industry index by at least *10 times*. If you invested $10,000 in a portfolio of the 10X companies at the end of 1972 (holding each enterprise at the general stock market rate of return until it came online on the New York Stock Exchange, the American Stock Exchange, or NASDAQ), your investment would have grown to be worth more than $6 million by the end of our study era (through 2002), a performance 32 times better than the general stock market.[2]

To grasp the essence of our study, consider one 10X case, Southwest Airlines. Just think of everything that slammed the airline industry from 1972 to 2002: Fuel shocks. Deregulation. Labor strife. Air-traffic-controller strikes. Crippling recessions. Interest-rate spikes. Hijackings. Bankruptcy after bankruptcy after bankruptcy. And in 2001, the terrorist attacks of September 11. And yet if you'd invested $10,000 in Southwest Airlines on December 31, 1972 (when it was just a tiny little outfit with three airplanes, barely reaching break-even and besieged by larger airlines out to kill the fledgling) your $10,000 would have grown to nearly $12 million by the end of 2002, a return 63 times better than the general stock market. It's a better performance than Wal-Mart, better than Intel, better than GE, better than Johnson & Johnson, better than Walt Disney. In fact, according to an analysis by *Money Magazine*, Southwest Airlines produced the #1 return to investors of all S&P 500 companies that were publicly traded in 1972 and held for a full 30 years to 2002.[3] These are impressive results by any measure, but they're astonishing when you take into account the roiling storms, destabilizing shocks, and chronic uncertainty of Southwest's environment.

Why did Southwest overcome the odds? What did it do to master its own fate? And how did it accomplish its world-beating performance when other airlines did not? Specifically, why did Southwest become great in such an extreme environment while its direct comparison, Pacific Southwest Airlines (PSA), flailed and was rendered irrelevant, despite having the same business model in the same industry with the same opportunity to become great? This single contrast captures the essence of our research question.

We've been asked by many of our students and readers, "How is this study different from your previous research into great companies, especially *Built to Last* and *Good to Great*?" The method is similar (comparative historical analysis) and the question of greatness is constant. But in this study, unlike any of the previous research, we selected cases not just on performance or stature but also on the extremity of the *environment*.

We selected on performance plus environment for two reasons. First, we believe the future will remain unpredictable and the world unstable for the rest of our lives, and we wanted to understand the factors that distinguish great organizations, those that prevail against extreme odds, in such environments. Second, by looking at the best companies and their leaders in extreme environments, we gain insights that might otherwise remain hidden when studying leaders in more tranquil settings. Imagine being on a leisurely hike, wandering along warm, sunlit meadows, and your companion is a great mountaineer who has led expeditions up the most treacherous peaks in the world. You'd probably notice that he's a little different from others, perhaps more watchful of the trail or more careful in packing his small daypack. But overall, given the safe predictability of a glorious spring day, it would be hard to see what really makes this leader so exceptional. Now, in contrast, envision yourself on the side of Mount Everest with this same climber, racing a murderous storm. In *that* environment, you'd see much more clearly what makes him different and what makes him great.

Studying leaders in an extreme environment is like conducting a behavioral-science experiment or using a laboratory centrifuge: throw leaders into an extreme environment, and it will separate the stark differences between greatness and mediocrity. Our study looks at how the truly great differed from the merely good in environments that exposed and amplified those differences.

In the remainder of this introductory chapter we briefly outline our research journey and preview some of the surprises we encountered along the way. (You can find a more detailed description of our research methodology in the *Research Foundations* appendices.) Starting in Chapter 2, we delve into what we learned about the individual people who led these companies, and in Chapters 3 through 6, how they led and built their companies differently from their less successful comparisons. In Chapter 7, we come to what, for us, was a particularly fascinating part of our journey: studying luck. We defined luck, quantified luck, determined if the 10X cases were luckier (or not), and discovered what they do *differently* about luck.

FINDING THE 10X CASES

We spent the first year of our efforts identifying the primary study set of 10X cases, searching for historical cases that met three basic tests:

1. The enterprise sustained truly spectacular results for an era of 15+ years relative to the general stock market and relative to its industry.
2. The enterprise achieved these results in a particularly turbulent environment, full of events that were uncontrollable, fast-moving, uncertain, and potentially harmful.
3. The enterprise began its rise to greatness from a position of vulnerability, being young and/or small at the start of its 10X journey.

From an initial list of 20,400 companies, we systematically sifted through 11 layers of cuts to identify cases that met all our tests. (See *Research Foundations: 10X-Company Selections*.) Because we wanted to study extreme performance in extreme environments, we used extreme standards in our selections. The final set of 10X cases (see the following table) delivered extraordinary performance during the dynastic eras we studied.

FINAL SET OF 10X CASES

10X Case	Dynastic Era of Study	Value of $10,000 Invested*	Performance Relative to Market[4]	Performance Relative to Industry[5]
Amgen	1980–2002	$4.5 million	24.0X the market	77.2X its industry
Biomet	1977–2002	$3.4 million	18.1X the market	11.2X its industry
Intel	1968–2002	$3.9 million	20.7X the market	46.3X its industry
Microsoft	1975–2002	$10.6 million	56.0X the market	118.8X its industry
Progressive Insurance	1965–2002	$2.7 million	14.6X the market	11.3X its industry
Southwest Airlines	1967–2002	$12.0 million	63.4X the market	550.4X its industry
Stryker	1977–2002	$5.3 million	28.0X the market	10.9X its industry

* Cumulative stock returns, dividends reinvested. Invest $10,000 in each company on December 31, 1972, and hold until December 31, 2002; if the company was not public on December 31, 1972, grow investment at general stock market rate of return until first month of CRSP data available for that company. Source for all stock-return calculations in this work: ©200601 CRSP®, Center for Research in Security Prices. Booth School of Business, The University of Chicago. Used with permission. All rights reserved. http://www.crsp.chicagobooth.edu.

Before we move on, let's address a key point about the cases in our study. We studied *historical eras* of dynastic performance that ended in 2002, not the companies as they are today. It's entirely possible that by the time you read these words, one or more of the companies on the list has stumbled, falling from greatness, leaving you to wonder, "But what about XYZ company? It doesn't seem to be a 10X performer today." Think of our research as comparable to studying a sports dynasty during its best years. Just because the UCLA Bruins basketball dynasty of the 1960s and 1970s under Coach John Wooden (with its 10 NCAA championships in 12 years) declined after Wooden retired does not invalidate insights obtained by studying the Bruins during its dynastic era.[6] In this same vein, a great company can cease to be great (see *How the Mighty Fall* by Jim Collins), yet this does not erase its dynastic era from the record books, and it's on that *historical dynastic era* that we focused our research lens and based our findings.

THE POWER OF CONTRAST

Our research method rests upon having a comparison set. The critical question is not "What did the great companies share in common?" The crucial question is "What did the great companies share in common that *distinguished* them from their direct comparisons?" Comparisons are companies that were in the same industry with the same or very similar opportunities during the same era as the 10X companies, yet that did not produce great performance. Using a rigorous scoring framework, we systematically identified a comparison company for each 10X case. (See *Research Foundations: Comparison-Company Selections.*) As a group, the 10X companies outperformed the comparison companies by more than 30 to 1 (see diagram "A Study In Contrasts").[7] The contrast between the 10X cases and the comparisons *during the relevant era of analysis* led to our findings.

Here then is the final study set of 10X cases and their comparisons: Amgen matched to Genentech; Biomet to Kirschner; Intel to AMD; Microsoft to Apple; Progressive to Safeco; Southwest Airlines to PSA; and Stryker to United States Surgical Corporation (USSC). Regarding the selection of Apple as a comparison case, we're aware that as of this writing in 2011, Apple stands as one of the most impressive comeback stories of all time. Our research lens for the Microsoft-versus-Apple contrast focused on the 1980s and 1990s, when Microsoft won big and Apple nearly killed itself. If you'd bought Apple stock at the end of December 1980, the month of its initial public offering (IPO), and held it to the end of our era of analysis in 2002, your investment would've ended up more than 80 percent *behind* the general stock market.[8] We'll address Apple's amazing resurgence under Steve Jobs later in this book, but one point is worth noting here: companies can indeed change over time, from comparison to 10X, and vice versa. It is always possible to go from good to great.

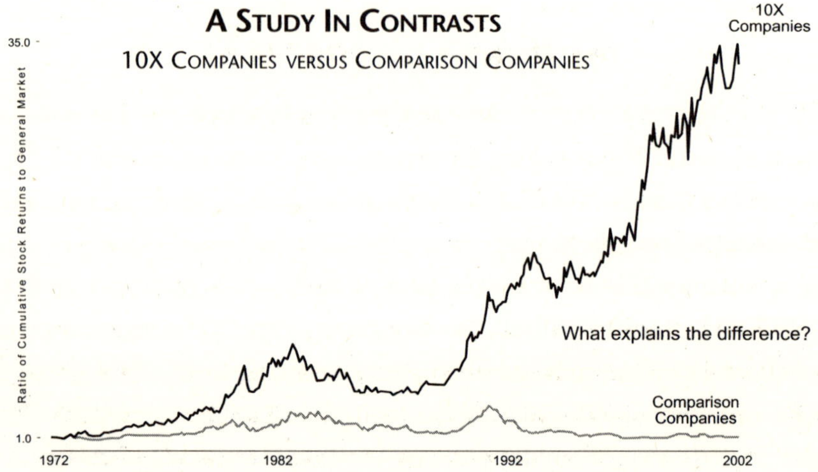

Notes:
1. Each company held at general stock market return until first month of first public CRSP data.
2. Source for all stock return calculations in this work: ©200601 CRSP®, Center for Research in Security Prices. Booth School of Business, The University of Chicago. Used with permission. All rights reserved. www.crsp.chicagobooth.edu.

SURPRISED BY THE DATA

We then performed a deep historical analysis of each pair of companies. We collected more than seven thousand historical documents to construct a clear understanding of how each company evolved, year by year, from founding through 2002. We systematically analyzed categories of data, including industry dynamics, founding roots, organization, leadership, culture, innovation, technology, risk, financial management, strategy, strategic change, speed, and luck. (See *Research Foundations* for more details on our data collection and analyses.) We didn't begin our journey with a theory to test or prove; we love being surprised by the evidence and changed by what we discover.

> We developed the concepts in this work from the data we gathered, building a framework from the ground up. We followed an iterative approach, generating ideas inspired by the data, testing those ideas against the evidence, watching them bend and buckle under the weight of evidence, replacing them with new ideas, revising, testing, revising yet again, until all the concepts squared with the evidence.

We placed the greatest weight on evidence from the actual time of the events. The core of our analysis always rested on comparing the 10X cases to the comparisons across time and asking, "What was different?" This method of inquiry proved particularly powerful for not only developing insights but also shattering deeply entrenched myths. In fact, many of the findings ran absolutely counter to our intuition and every major finding surprised at least one of us. As a preview of what's to come, here is a sampling of myths undermined by the research.

Entrenched myth: Successful leaders in a turbulent world are bold, risk-seeking visionaries.
Contrary finding: The best leaders we studied did not have a visionary ability to predict the future. They observed what worked, figured out *why* it worked, and built upon proven foundations. They were not more risk taking, more bold, more visionary, and more creative than the comparisons. They were more disciplined, more empirical, and more paranoid.

Entrenched myth: Innovation distinguishes 10X companies in a fast-moving, uncertain, and chaotic world.
Contrary finding: To our surprise, no. Yes, the 10X cases innovated, a lot. But the evidence does not support the premise that 10X companies will necessarily be more innovative than their less successful comparisons; and in some surprise cases, the 10X cases were *less* innovative.

Innovation by itself turns out not to be the trump card we expected; more important is the ability to *scale* innovation, to blend creativity with discipline.

Entrenched myth: A threat-filled world favors the speedy; you're either the quick or the dead.
Contrary finding: The idea that leading in a "fast world" always requires "fast decisions" and "fast action"—and that we should embrace an overall ethos of "Fast! Fast! Fast!"—is a good way to get killed. 10X leaders figure out *when* to go fast, and when *not* to.

Entrenched myth: Radical change on the outside requires radical change on the inside.
Contrary finding: The 10X cases changed *less* in reaction to their changing world than the comparison cases. Just because your environment is rocked by dramatic change does not mean that you should inflict radical change upon yourself.

Entrenched myth: Great enterprises with 10X success have a lot more good luck.
Contrary finding: The 10X companies did not generally have more luck than the comparisons. Both sets had luck—*lots* of luck, both good and bad—in comparable amounts. The critical question is not whether you'll have luck, but what you *do* with the luck that you get.

A NEW LENS, AN ENDURING QUEST

This book adds to a body of work on what separates great companies from good ones that began in 1989 with the *Built to Last* research (conducted with Jerry Porras), and continued with the *Good to Great* research and the *How the Mighty Fall* analysis. The complete data set from all this research covers the evolution of 75 corporations, for a total of more than six thousand years of combined corporate history.[9] So, while this is a distinctive and original piece of research, it can also be seen as an integral part of a longer journey to explore one question, "What does it take to build a great company?"

We think of each research study as like punching holes and shining a light into a black box, inside which we find enduring principles that distinguish great companies from good ones. Each new study uncovers additional dynamics and allows us to see previously discovered principles from new angles. We cannot claim that the concepts we uncover "cause" greatness (no one in the social sciences can ever claim causality), but we can claim correlations rooted in the evidence. If you apply our findings with discipline, your chances of building an enduring great company will be higher than if you behave like a comparison case.

If you've read *Built to Last, Good to Great,* or *How the Mighty Fall,* you'll notice very little discussion in the next six chapters about the concepts uncovered in those works. With the exception of a direct link to Level 5 leadership, we've deliberately *not* written in the coming pages about principles like the Hedgehog Concept, First Who (the right people on the bus), core values, BHAGs (Big Hairy Audacious Goals), cult-like cultures, the Stockdale Paradox, clock building, the five stages of decline, or the flywheel. The reason is simple: why dwell on what's already well covered in the previous books in this book? That said, we did test the principles from the previous books and found that they *do* apply in a chaotic and uncertain world. At the end of this book (see *Frequently Asked Questions*), we'll address common questions about how the concepts in this work link to those in prior books. But the primary purpose of this book is to share the *new* concepts learned from *this* study.

Now that we've completed our research journey, we feel a much greater sense of calm. Not because we believe life will magically become stable and predictable; if anything, the forces of complexity, globalization, and technology are accelerating change and increasing volatility. We feel calm because we have increased understanding of what it takes to survive, navigate, and prevail. We are much better prepared for what we cannot possibly predict.

Thriving in a chaotic world is not just a business challenge. In fact, all our work is not fundamentally *about* business, but about the principles that distinguish great organizations from good ones. We're curious to discover what makes for enduring great organizations of *any* type. We use publicly traded corporations as the data set because they provide a clear and consistent metric of results (so we can carefully select our study cases), and easily accessible and extensive historical data. A great public school, a great hospital, a great sports team, a great church, a great military unit, a great homeless shelter, a great orchestra, a great non-profit—each has its own definition of results, defined by its core purpose—yet the question of what it takes to achieve superior performance amidst unrelenting uncertainty faces them all. Greatness is not just a business quest; it's a human quest.

So, we invite you to join us on a journey to learn what we learned. Challenge and question; let the evidence speak. Take what you find useful and apply it to creating a great enterprise that doesn't just react to events but shapes events. As the influential management thinker Peter Drucker taught, the best—perhaps even the only—way to predict the future is to create it.[10]

2

10XERS

"Victory awaits him who has everything in order—luck people call it. Defeat is certain for him who has neglected to take the necessary precautions in time; this is called bad luck."

—Roald Amundsen, *The South Pole*[1]

In October 1911, two teams of adventurers made their final preparations in their quest to be the first people in modern history to reach the South Pole. For one team, it would be a race to victory and a safe return home. For members of the second team, it would be a devastating defeat, reaching the Pole only to find the wind-whipped flags of their rivals planted 34 days earlier, followed by a race for their lives—a race that they lost in the end, as the advancing winter swallowed them up. All five members of the second Pole team perished, staggering from exhaustion, suffering the dead-black pain of frostbite and then freezing to death as some wrote their final journal entries and notes to loved ones back home.

It's a near-perfect matched pair. Here we have two expedition leaders—Roald Amundsen, the winner, and Robert Falcon Scott, the loser—of similar ages (39 and 43) and with comparable experience. Amundsen led the first successful journey through the Northwest

Passage and joined the first expedition to spend the winter in Antarctica; Scott led a South Pole expedition in 1902, reaching 82 degrees South. Amundsen and Scott started their respective journeys for the Pole within days of each other, both facing a round trip of more than fourteen hundred miles (roughly equal to the distance from New York City to Chicago and back) into an uncertain and unforgiving environment, where temperatures could easily reach 20 degrees below zero F even during the summer, made worse by gale-force winds. And keep in mind, this was 1911. They had no means of modern communication to call back to base camp—no radio, no cell phones, no satellite links—and a rescue would have been highly improbable at the South Pole if they screwed up. One leader led his team to victory and safety. The other led his team to defeat and death.[2]

What separated these two men? Why did one achieve spectacular success in such an extreme set of conditions, while the other failed even to survive? It's a fascinating question and a vivid analogy for our overall topic. Here we have two leaders, both on quests for extreme achievement in an extreme environment. And it turns out that the 10X business leaders in our research behaved very much like Amundsen and the comparison leaders behaved much more like Scott. We'll turn to the business leaders in a few pages, but first let's add a bit more detail to the tale of Amundsen and Scott. (To learn even more about Amundsen and Scott, we recommend starting with Roland Huntford's superb book *The Last Place on Earth*, a massive, well-written comparative study of these two men.)

ARE YOU AMUNDSEN OR SCOTT?

While in his late twenties, Roald Amundsen traveled from Norway to Spain for a two-month sailing trip to earn a master's certificate. It was 1899. He had a nearly two-thousand-mile journey ahead of him. And how did Amundsen make the journey? By carriage? By horse? By ship? By rail?

He bicycled.

Amundsen then experimented with eating raw dolphin meat to determine its usefulness as an energy supply. After all, he reasoned, someday he might be shipwrecked, finding himself surrounded by dolphins, so he might as well know if he could eat one.

It was all part of Amundsen's years of building a foundation for his quest, training his body and learning as much as possible from practical experience about what actually worked. Amundsen even made a pilgrimage to apprentice with Eskimos. What better way to learn what worked in polar conditions than to spend time with a people who have hundreds of years of accumulated experience in ice and cold and snow and wind? He learned how Eskimos used dogs to pull sleds. He observed how Eskimos never hurried, moving slowly and steadily, avoiding excessive sweat that could turn to ice in sub-zero temperatures. He adopted Eskimo clothing, loose fitting (to help sweat evaporate) and protective. He systematically practiced Eskimo methods and trained himself for every conceivable situation he might encounter en route to the Pole.

Amundsen's philosophy: You don't wait until you're in an unexpected storm to discover that you need more strength and endurance. You don't wait until you're shipwrecked to determine if you can eat raw dolphin. You don't wait until you're on the Antarctic journey to become a superb skier and dog handler. You prepare with intensity, all the time, so that when conditions turn against you, you can draw from a deep reservoir of strength. And equally, you prepare so that when conditions turn in your favor, you can strike hard.

Robert Falcon Scott presents quite a contrast to Amundsen. In the years leading up to the race for the South Pole, he could have trained like a maniac on cross-country skis and taken a thousand-mile bike ride. He did not. He could have gone to live with Eskimos. He did not. He could have practiced more with dogs, making himself comfortable with choosing dogs over ponies. Ponies, unlike dogs, sweat on their hides so they become encased in ice sheets when tethered, posthole and struggle in snow, and don't generally eat meat. (Amundsen planned to kill some of the weaker dogs along the way to fuel the stronger dogs.)

Scott chose ponies. Scott also bet on "motor sledges" that hadn't been fully tested in the most extreme South Pole conditions. As it turned out, the motor-sledge engines cracked within the first few days, the ponies failed early, and his team slogged through most of the journey by "man-hauling," harnessing themselves to sleds, trudging across the snow, and pulling the sleds behind them.

Unlike Scott, Amundsen systematically built enormous buffers for unforeseen events. When setting supply depots, Amundsen not only flagged a primary depot, he placed 20 black pennants (easy to see against the white snow) in precise increments for miles on either side, giving himself a target more than ten kilometers wide in case he got slightly off course coming back in a storm. To accelerate segments of his return journey, he marked his path every quarter of a mile with packing-case remnants and every eight miles with black flags hoisted upon bamboo poles. Scott, in contrast, put a single flag on his primary depot and left no markings on his path, leaving him exposed to catastrophe if he went even a bit off course. Amundsen stored three tons of supplies for 5 men starting out versus Scott's one ton for 17 men. In his final push for the South Pole from 82 degrees, Amundsen carried enough extra supplies to miss every single depot and still have enough left over to go another hundred miles. Scott ran everything dangerously close to his calculations, so that missing even one supply depot would bring disaster. A single detail aptly highlights the difference in their approaches: Scott brought one thermometer for a key altitude-measurement device, and he exploded in "an outburst of wrath and consequence" when it broke; Amundsen brought *four* such thermometers to cover for accidents.

Amundsen didn't know precisely what lay ahead. He didn't know the exact terrain, the altitude of the mountain passes, or all the barriers he might encounter. He and his team might get pounded by a series of unfortunate events. Yet he designed the entire journey to systematically reduce the role of big forces and chance events by vigorously embracing the possibility of those very same big forces and chance events. He *presumed* bad events might strike his team somewhere along the journey and he prepared for them, even developing contingency plans so that the team could go on should something unfortunate happen to him along the way. Scott left himself unprepared and complained in his journal about his bad luck. "Our luck in weather is preposterous," penned Scott in his journal, and wrote in another entry, "It is more than our share of ill-fortune . . . How great may be the element of luck!"

On December 15, 1911, in bright sunshine sparkling across the vast white plain, with a slight crosswind and a temperature of 10 degrees below zero F, Amundsen reached the South Pole. He and his teammates planted the Norwegian flag, which "unfurled itself with a sharp crack," and dedicated the plateau to the Norwegian king. Then they went right back to work. They erected a tent and attached a letter to the Norwegian king describing their success; Amundsen addressed the envelope to Captain Scott (presuming Scott would be the next to reach the Pole) as an insurance policy in case his team met an unfortunate end on the journey home. He could not have known that Scott and his team were man-hauling their sleds, fully 360 miles behind.

More than a month later, at 6:30 p.m. on January 17, 1912, Scott found himself staring at Amundsen's Norwegian flag at the South Pole. "We have had a horrible day," Scott wrote in his diary. "Add to our disappointment a head wind 4 to 5, with a temperature –22° . . . Great God! this is an awful place and terrible enough for us to have labored to it without the reward of priority." On that very day, Amundsen had already traveled nearly five hundred miles back north, reaching his 82-degree supply depot with only eight easy days to go. Scott turned around and headed back north, more than seven hundred miles of man-hauling from home base, just as the season began to turn. The weather became more severe, with increasing winds and decreasing temperatures, while supplies dwindled and the men struggled through the snow.

Amundsen and his team reached home base in good shape on January 25, the precise day he'd penned into his plan. Running out of supplies, Scott stalled in mid-March, exhausted and depressed. Eight months later, a British reconnaissance party found the frozen bodies of

Scott and two companions in a forlorn, snow-drifted little tent, just ten miles short of his supply depot.[3]

DIFFERENT BEHAVIORS, NOT DIFFERENT CIRCUMSTANCES

Amundsen and Scott achieved dramatically different outcomes *not* because they faced dramatically different circumstances. In the first 34 days of their respective expeditions, Amundsen and Scott had *exactly* the same ratio, 56 percent, of good days to bad days of weather.[4] If they faced the same environment in the same year with the same goal, the causes of their respective success and failure simply cannot be the environment. They had divergent outcomes principally because they displayed very *different behaviors*.

So too, with the leaders in our research study. Like Amundsen and Scott, our matched pairs were vulnerable to the same environments at the same time. Yet some leaders proved themselves to be 10Xers while leaders on the other side of the pair did not. "10Xers" (pronounced "ten-EX-ers") is our term for the people who built the 10X companies. In our research, we observed that the 10Xers shared a set of behavioral traits that distinguished them from the comparison leaders. In this chapter we introduce these traits, and in subsequent chapters we describe how our 10Xers led and built their successful companies consistent with them.

Let's first look at what we did *not* find about 10Xers relative to their less successful comparisons.

They're *not* more creative.
They're *not* more visionary.
They're *not* more charismatic.
They're *not* more ambitious.
They're *not* more blessed by luck.
They're *not* more risk seeking.
They're *not* more heroic.
They're *not* more prone to making big, bold moves.

To be clear, we're not saying that 10Xers lacked creative intensity, ferocious ambition, or the courage to bet big. They displayed all these traits, but *so did their less successful comparisons.*

So then, how did the 10Xers distinguish themselves? First, 10Xers embrace a paradox of control and non-control.

> On the one hand, 10Xers understand that they face continuous uncertainty and that they cannot control, and cannot accurately predict, significant aspects of the world around them. On the other hand, 10Xers reject the idea that forces outside their control or chance events will determine their results; they accept full responsibility for their own fate.

10Xers then bring this idea to life by a triad of core behaviors: *fanatic discipline, empirical creativity,* and *productive paranoia.* Animating these three core behaviors is a central motivating force, *Level 5 ambition.* (See diagram "10X Leadership.") These behavioral traits, which we introduce in the remainder of this chapter, correlate with achieving 10X results in chaotic and uncertain environments. Fanatic dis-

10X LEADERSHIP

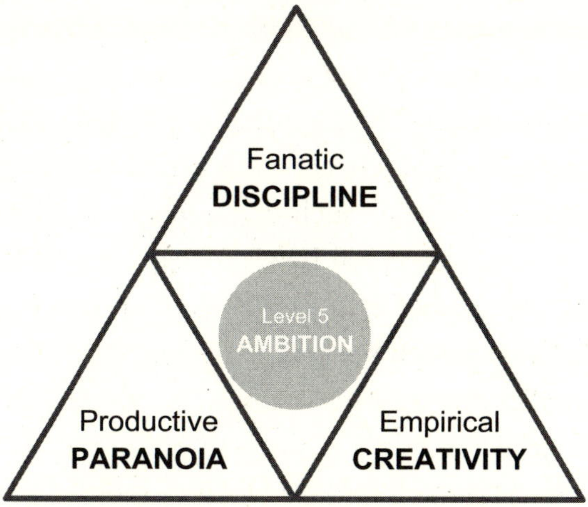

cipline keeps 10X enterprises on track, empirical creativity keeps them vibrant, productive paranoia keeps them alive, and Level 5 ambition provides inspired motivation.

FANATIC DISCIPLINE

In the late 1990s, Peter Lewis, CEO of Progressive Insurance, faced a seemingly irrational Wall Street driving Progressive's stock price wildly up and down. On October 16, 1998, Progressive's stock jumped nearly $20, an 18 percent jump in a single day. Did anything fundamentally change about the company that day? No. Did the economy make a sudden lurch? No. Did the market rally 18 percent that day? No. Absolutely nothing of any significance had changed for Progressive on October 16, 1998. Yet the stock price soared an astounding 18 percent.

Then in the very next quarter, on January 26, 1999, Progressive's stock plummeted nearly $30, a 19 percent drop in a single day. Did anything fundamentally change about the company that day? No. Did the economy make a sudden lurch? No. Did the market crash? No. Absolutely nothing of any significance had changed for Progressive on January 26, 1999. Yet the stock price fell an astounding 19 percent.[5]

These fluctuations stemmed in part from Peter Lewis's belief that playing earnings games to satisfy Wall Street lacked honesty. He refused to play the game of telling analysts about forthcoming earnings so that they could more reliably "predict" those very same earnings, a behavior Lewis saw as a shortcut alternative to deep analysis and field work. Lewis also rejected the idea that a company should "manage earnings" by smoothing them out from quarter to quarter so as not to rattle the markets, viewing such shenanigans as undisciplined. But this caused a problem. Because Lewis rejected the "I'll tell you what we'll earn and you predict what we'll earn and we'll both be happy" model, and because he refused to smooth earnings, analysts couldn't consistently predict Progressive's earnings. As one analyst complained, "I might as well flip a coin."[6]

And so, on October 16, 1998, Progressive exceeded analyst expectations by 44 cents a share, driving the stock up, and then on January 26, 1999, Progressive's earnings fell below analyst expectations by 16 cents a share, driving the stock down. If Lewis were to continue to refuse to play the game, Progressive's stock price would continue to spike up and down, which could make the company vulnerable to raiders. To ignore that risk would be like a polar explorer choosing to ignore the possibility of a freak storm that could kill him. Yet capitulating would compromise Lewis's principles. What was Lewis to do?

He rejected Option A (to ignore) and Option B (to capitulate), and chose Option Q. Progressive would become the first SEC-listed company to publish *monthly* financial statements. This would give analysts actual performance data as the quarter progressed, from which they could more easily estimate quarterly results. Other companies had capitulated to the guidance game because, well, they felt they had no choice, that they were imprisoned by this huge force out of their control. But Lewis freed Progressive from the prison. He accepted that these pressures existed, yet he mitigated their effect by prodigious effort.[7]

What does this story have to do with "discipline"?

Discipline, in essence, is *consistency of action*—consistency with values, consistency with long-term goals, consistency with performance standards, consistency of method, consistency over time. Discipline is not the same as regimentation. Discipline is not the same as measurement. Discipline is not the same as hierarchical obedience or adherence to bureaucratic rules. True discipline requires the independence of mind to reject pressures to conform in ways incompatible with values, performance standards, and long-term aspirations. For a 10Xer, the only legitimate form of discipline is self-discipline, having the inner will to do whatever it takes to create a great outcome, no matter how difficult.

10Xers are utterly relentless, monomaniacal even, unbending in their focus on their quests. They don't overreact to events, succumb to the herd, or leap for alluring—but irrelevant—opportunities. They're capable of immense perseverance, unyielding in their standards yet disciplined enough not to overreach. In our research-team discussions, we struggled with how to best describe the discipline we found in the 10X leaders. Most business CEOs have some level of discipline, but the 10Xers operated on an entirely different level. The 10Xers, we concluded, weren't just disciplined; they were *fanatics*. Lewis's decision to issue monthly financial reports is akin to Amundsen's riding his bicycle from Norway to Spain and eating raw dolphin meat; their behavior fits nowhere on a normal curve.

Herb Kelleher of Southwest Airlines believed passionately in sustaining a high-spirit, fun-loving, and iconoclastic culture full of passionate people infused with a rebellious "Warrior Spirit."[8] Kelleher understood that superb customer service naturally arises when people have fun at work and love their company. As the airline grew from a small Texas commuter airline with only a handful of airplanes into a major national carrier, it would be increasingly difficult, and increasingly important, to sustain the culture. So, Kelleher himself behaved as a fanatic exemplar of the culture.

"I will bet you one thing," Kelleher told *60 Minutes*, "that I'm the only airline president in America that would go over to his maintenance hangar at two o'clock in the morning in a flowered hat with a feathered boa and a purple dress." [9] When asked to grace the cover of *Texas Monthly* magazine, he showed up in a white suit, zipped down to show off his bare chest; the cover shot portrayed him doing some sort of an Elvis-like dance next to the headline "Herbie Goes Bananas."[10] When he faced a trade-slogan-ownership dispute with Stevens Aviation, he met Stevens's CEO not in the courtroom, but in an arena filled with hundreds of employees punching the air with pompoms—to resolve the matter with an arm-wrestling contest.[11] We on the research team joked that Kelleher's Technicolor quirks evoked a Hunter S. Thompson quote with a slight twist: when the going gets weird, the weird become CEO.

But to focus on Kelleher's weirdness *as* weirdness would miss the point. He wasn't weird to be weird; he was behaving with outlandish consistency to animate the culture, like an impactful actor who stays perfectly in character while on stage. He was also a complete monomaniac about building Southwest Airlines, never resting in the quest to make Southwest the best low-cost, high-spirit airline, winning every battle and every war with its competitors. "In my spare time, I work," Kelleher explained in 1987, "seven days a week, usually until 8 or 9 o'clock at night," then he'd settle down before bed to make progress on reading the thousands of books scattered about his home.[12] Kelleher was like Muhammad Ali, combining a deadly serious intensity with a blustery, comical exterior. You might laugh with Kelleher, much like enjoying an Ali press conference, but then find yourself flat on your back if you dared to square off in the ring. By one account, Kelleher showed his competitive ferocity speaking to a gathering of Southwest people, "If someone says they're going to smack us in the face—knock them out, stomp them out, boot them in the ditch, cover them over and move on to the next thing."[13]

> Both Kelleher and Lewis, like all the 10Xers we studied, were nonconformists in the best sense. They started with values, purpose, long-term goals, and severe performance standards; and they had the fanatic discipline to adhere to them. If that required them to diverge from normal behavior, then so be it. They didn't let external pressures, or even social norms, knock them off course. In an uncertain and unforgiving environment, following the madness of crowds is a good way to get killed.

And why would they have such independence of mind? Not because they had more inherent audacity than others, and not because they were more brash and rebellious than others, but because they were more *empirical*, which brings us to the second of the three core 10Xer behaviors.

EMPIRICAL CREATIVITY

In 1994, Andy Grove, chief executive of Intel, underwent a routine blood test that came back with a worrisome number: a PSA (prostate-specific antigen) reading of 5, indicating that there could be a tumor the size of a sugar cube growing inside his prostate gland. The doctor suggested that Grove's first step should be to visit the urologist. Most people would do exactly that, but that wasn't Andy Grove's response. Instead, he began reading research articles written *by* medical scientists *for* medical scientists. Grove delved into the data. What did the PSA test really indicate? How did the biochemistry work? What were the statistics of prostate cancer, and the pros and cons of treatment options? He also decided to "test the tests" to validate the data in his readings, sending blood samples to separate labs to calibrate the degree of lab variation in the test. Only after all this did Grove make an appointment with the urologist.

But even then, Grove did not rely on his doctors to create a treatment plan. After an MRI and a bone scan, he embarked on a more extensive research regimen, going directly to original sources, culling through the primary data. He obtained all the articles cited in the bibliography of a prostate-cancer reference book, devoured those, then searched for scientific literature that had been published in the six to nine months after the publication of that book, and then obtained even *more* materials that'd been cited in those publications. Grove maintained an intense CEO schedule by day and his prostate research regimen by night, plotting data, cross-referencing different studies, and trying to make sense of it all. He learned through his research that there was a raging intellectual war over various cancer-treatment regimens. Grove realized he ultimately had to draw his own decision trees; plug in his own probability equations; and come to his own data-driven, logical conclusions about his treatment plan. "As a patient whose life and well-being depended on a meeting of minds," he later wrote in *Fortune* magazine, "I realized I would have to do some cross-disciplinary work on my own."[14]

After electing to undergo a biopsy, which confirmed the presence of a moderately aggressive tumor, Grove threw his prodigious mental capacity at the question of what he should do next. Cancer treatments usually involve some combination of slicing you up (surgery), frying you (radiation), or poisoning you (chemotherapy); and each option has its own side effects, consequences, and survival rates. Furthermore, each doctor tends to have a bias toward a particular treatment, influenced by that doctor's own specialties (if you're a hammer, everything you see looks like a nail). Grove found proponents of traditional surgery, cryosurgery, external radiation, seed therapy, high-dose-rate radiation, and combination therapies. The dominant conventional wisdom pointed to surgery, but Grove's *own* direct engagement with the evidence led him to a different choice (a combination radiation therapy). In the end, Grove reflected, "I decided to bet on my own charts."[15]

Now, you might be thinking, "My goodness, what an arrogant jerk! Who does he think he is to defy the whole medical establishment?" But think about it this way: Grove discovered that the medical establishment itself had great uncertainty and disagreement within its own ranks, a dynamic amplified by rapidly advancing technologies. Had Grove faced a broken arm, with no uncertainty about treatment and zero risk of death, he wouldn't have spent hundreds of hours building charts of data. But with significant uncertainty multiplied by significant consequences, Grove did what all our 10Xers did, he turned directly to empirical evidence.

> Social psychology research indicates that at times of uncertainty, most people look to other people—authority figures, peers, group norms—for their primary cues about how to proceed.[16] 10Xers, in contrast, do not look to conventional wisdom to set their course during times of uncertainty, nor do they primarily look to what other people do, or to what pundits and experts say they should do. They look primarily to empirical evidence.

The point here is not to be contrary and independent just for the sake of being contrary and independent. The point is to be more empirical to buttress your mental independence and validate your creative instincts. By "empirical," we mean relying upon direct observation, conducting practical experiments, and/or engaging directly with evidence rather than relying upon opinion, whim, conventional wisdom, authority, or untested ideas. Having an empirical foundation enables 10Xers to make bold, creative moves *and* bound their risk. Andy Grove's approach to his cancer treatment was unusual, even creative, yet deeply grounded in evidence and rigor.

In planning for the South Pole expedition, Amundsen set up his base camp in a location no one else had seriously considered, a bold stroke that put him sixty miles closer to the South Pole from the get-go. Everyone believed McMurdo Sound was the best place to launch a bid for the Pole. It had been used by other explorers and had proven to be a stable place to build a base. But Amundsen saw another option, the Bay of Whales. Other expedition leaders believed the Bay of Whales to be unstable ice and thereby a foolhardy place to base operations. Amundsen gathered the source notes and journals from previous expeditions, dating back to Ross's voyage in 1841. He pored over the details, immersing himself in the evidence, noting consistencies and discrepancies, and assessing all the options. He noticed something interesting, something missed by others who simply accepted the conventional distrust of the Bay of Whales: a dome-like feature that'd remained in the same place for seven decades. Amundsen concluded that this particular part of the barrier was in fact a stable location. Wrote Huntford of this decision, "Amundsen was the first to draw the obvious conclusion because he was the first to study the sources . . . [He] was that rare creature, an intellectual Polar explorer; with the capacity to examine evidence and make logical deductions."[17]

> The 10Xers did not generally make bolder moves than their less successful comparisons; both groups made big bets and, when needed, took dramatic action. Nor did the 10Xers exude more raw confidence than the comparison leaders; indeed, the comparison leaders were often brazenly self-confident. But the 10Xers had a much deeper empirical foundation for their decisions and actions, which gave them well-founded confidence and bounded their risk.

Does all of this emphasis on being empirical make 10Xers indecisive? Not really. Grove took decisive action on his cancer once he'd immersed himself in the evidence, just as Amundsen took decisive action to land at the Bay of Whales. The 10Xers don't favor analysis over action; they favor *empiricism as the foundation* for decisive action.

Yet despite their empirical confidence, 10Xers never feel safe or comfortable; indeed, they remain afraid—terrified, even—of what the world can throw at them. So, they prepare to meet head-on what they most fear, which brings us to the third core behavior.

PRODUCTIVE PARANOIA

In early 1986, Microsoft leaders met with underwriters and lawyers to edit the prospectus for an initial public stock offering. The underwriters and lawyers came prepared to be the purveyors of darkness, to engage in a battle with Microsoft leaders to adequately describe the risks investors should consider. But instead of encountering an overly optimistic entrepreneurial leader who painted a rosy picture of unstoppable success, they met DOCTOR DOOM. Steve Ballmer, then a vice president, reveled in coming up with scenario after scenario of risk, peril, danger, death, crippling attack, misfortune, and catastrophe. Grim possibilities poured into the conversation, underwriters scribbling away. Finally, after pausing to digest all the possible carnage, one of the underwriters said to Ballmer, "I'd hate to hear you on a bad day."[18]

Ballmer became the Commissar of Concern under tutelage from the Grand Master of Productive Paranoia himself, Bill Gates. Ballmer had abandoned his studies at the Stanford Graduate School of Business to join his friend's adventure. As Ballmer recalled, he did some calculations about growth and concluded that Microsoft needed to hire 17 people. Gates threw a fit. Seventeen people? Did Ballmer want to bankrupt the company? Seventeen people? No way! Seventeen people? Microsoft would never expose itself to financial ruin! Seventeen people? Microsoft should have enough cash on hand to go a year—an entire year!—without a penny of revenues.[19]

"Fear should guide you, but it should be latent," Gates said in 1994. "I consider failure on a regular basis." He hung a photograph of Henry Ford in his office, to remind himself that even the greatest entrepreneurial successes can be passed by, as Ford had been passed by GM in the early history of the auto industry. He worried constantly about who might be the next Bill Gates, some freaky high school kid toiling away 22 hours a day in some dingy little office coming up with a lethal torpedo to fire at Microsoft.[20]

Gates showed his fearful side in what became known as the "nightmare memo." In a four-day period, from June 17 to June 20, 1991, Bill Gates's personal fortune dropped more than $300 million as Microsoft stock suddenly fell 11 percent when a memo filled with "nightmare" scenarios leaked its way to the *San Jose Mercury News*. Written by Gates himself, the memo listed a series of worries and threats—about competitors, technology, intellectual property, legal cases, and Microsoft's customer-support shortcomings—and proclaimed that "our nightmare . . . is a reality." Keep in mind that at the time of the memo, Microsoft was rapidly becoming the most powerful player in its industry, with Windows on the verge of becoming one of the most dominant software products ever. Anyone who understood Gates would've known that the memo didn't signal a change; he'd *always* lived in fear, always felt vulnerable, and he would continue to do so. "If I really believed this stuff about our invincibility," he said the year after the nightmare memo, "I suppose I would take more vacations."[21]

Quite a contrast to John Sculley, who presided over Apple during much of its comparison era in the mid-1980s to the early 1990s. In 1988, Apple had a spectacularly good year. *USA Today* reported, "Apple isn't just on the rebound—it's bounding ahead faster than it has since 1983. In each of the past three quarters, revenues climbed more than 50% above the same year-ago period, while net income shot up more than 100%. At this rate, the computer maker will finish 1988 doubling both sales and net income in just two years." And how did Sculley respond? Did he live in fear that Apple's very success might presage possible doom?

He announced a nine-week sabbatical.[22]

Nine weeks!

To be fair, Sculley didn't plan to disappear entirely for nine weeks; he'd still attend board meetings, meet with securities analysts, and appear at MacWorld, among other activities. Still, it's quite a contrast to Gates's responding to success by worrying obsessively and issuing nightmare memos. The same *USA Today* article quoted Sculley, "I've got the team in place here. Things are booming. So I'm going fishing."[23]

The very next year, Apple's return on equity began to fall, from nearly 40 percent in 1988 to 13 percent in 1994 (Sculley had left Apple by this point) and turning *negative* in 1996. Apple continued to hurtle downward until Steve Jobs's return in the late 1990s.[24] Our point is not that a sabbatical caused Apple's decline or that John Sculley was lazy (when fully engaged, he had a prodigious work ethic); our point is to draw a contrast with the productive paranoia Gates demonstrated *all the time*, no matter how successful Microsoft became.[25]

> 10Xers differ from their less successful comparisons in how they maintain hypervigilance in good times as well as bad. Even in calm, clear, positive conditions, 10Xers constantly consider the possibility that events could turn against them at any moment. Indeed, they believe that conditions will—absolutely, with 100 percent certainty—turn against them without warning, at some unpredictable point in time, at some highly inconvenient moment. And they'd better be prepared.

Whether it be Herb Kelleher at Southwest Airlines predicting 11 of the last 3 recessions, Andy Grove of Intel "looking for the black cloud in the silver lining," Kevin Sharer of Amgen putting a portrait of General George A. Custer (who led his troops to calamity at Little Big Horn) in his office to remind himself that overconfidence leads to doom, or Bill Gates issuing nightmare memos at Microsoft, the 10Xers have a consistent pattern. By embracing the myriad of possible dangers, they put themselves in a superior position to overcome danger.[26]

10Xers distinguish themselves not by paranoia per se, but by how they take *effective action* as a result. Paranoid behavior is enormously functional *if* fear is channeled into extensive preparation and calm, clearheaded action, hence our term "*productive* paranoia." (We're not making any claims about clinical paranoia here; we're labeling instead the 10X behavior of turning hypervigilance into preparation and productive action.) Gates didn't just sit around writing up nightmare memos; he channeled fear into action by keeping workspace inexpensive; hiring better people; building cash reserves; and working on the next software release to stay a step ahead, then the next one, and the next one after that. Like Amundsen with his huge supply buffers, 10Xers maintain a conservative financial position, squirreling away cash to protect against unforeseen disruptions. Like Amundsen sensing great risk in betting on unproven methods and technologies, they avoid unnecessary risks that could expose them to calamity. Like Amundsen, they succeed in an uncertain and unforgiving environment through deliberate, methodical, and systematic preparation, always asking, "What if? What if? What if?"

Productive paranoia isn't just about avoiding danger, trying to find the safest and most enjoyable path through life; 10Xers seek to accomplish a great objective, be it a goal, a company, a noble ambition to change the world, or a desire to be useful in the extreme. Indeed, as an overall life approach, they worry not about protecting what they have, but creating and building something truly great, bigger than themselves, which brings us to the motivating force behind the three core 10Xer behaviors.

LEVEL 5 AMBITION

At first, we wondered, "Why would anyone work with these people?" They seem, well, somewhat extreme: paranoid, contrarian, independent, obsessed, monomaniacal, exhausting, and so forth. Early in our research conversations, we labeled them PNFs, short for "paranoid, neurotic freaks." Yet the fact is, they attracted thousands of people to join them in their respective quests. If they were nothing but weird, selfish, antisocial, paranoid freaks of nature, they likely could not have built truly great organizations. So, why did people follow them? Because of a deeply attractive form of ambition: 10Xers channel their ego and intensity into something larger and more enduring than themselves. They're ambitious, to be sure, but for a purpose beyond themselves, be it building a great company, changing the world, or achieving some great object that's ultimately not about them.

In 1992, *Business Week* published a special report on the relationship between CEO pay and corporate performance. Dane Miller of Biomet (one of the 10X companies in our study) ranked #1, delivering more value per dollar of his own compensation than any other CEO. And it wasn't just a one-year blip. He sustained a top ranking—sometimes #1, always near the top—for more than a decade in publications like *Forbes*, *Business Week*, and *Chief Executive Magazine*. Keep in mind, the 1990s became the acceleration point when executive compensation began spiraling upward, fueled by stock options that gave CEOs massive upside if their companies did well but minimal downside if their companies fared poorly. Miller's stock-option package at the time? Zero. His employees had options but he did not. He owned his own equity outright so that his personal fortune linked directly to the company's performance on the upside *and* the downside.[27] In a sense, relative to business norms, Miller could have been viewed as the world's most underpaid CEO.

Yet Miller showed nothing but gratitude, noting in 2000 that his life was dedicated to two things, Biomet and his family. "There's nothing else I want to do in my life," said Miller. "I enjoy virtually every day and I couldn't be having any more fun or any more excitement about what I do." As for being the most underpaid CEO relative to value, he blasted the idea of granting tons of upside-only options. What's the point of just more and more and more for the sake of more and more and more? "What incremental value does an extra 100,000 shares have?" he snorted. "At some point, you're just satisfying an uncontrollable greed complex."[28]

In *Good to Great*, we wrote about Level 5 leaders, those who lead with a powerful mixture of personal humility plus professional will. Every good-to-great transition in that research began with the emergence of a Level 5 leader who deflected attention from himself, maintained a low profile, and led with inspired standards rather than inspiring personality. On the surface, some of the 10Xers appear to be unlike Level 5 leaders. Kelleher had a zany and flamboyant personality who often drew attention to himself by his antics. So did Peter Lewis. In culling through decades of documents on the Lewis era at Progressive Insurance, we came across a range of descriptors: "Just plain strange." "Oddball." "A standard deviation from an iconoclast." "A Wildman." "Eccentric." "A frame or two off the ordinary screen." "A rock star without any musical ability." "No way to jerk his chain because he doesn't have one."[29] Lewis signed his annual letter to shareholders with the quirky "Joy, Love and Peace—Peter B. Lewis." He strode into a board meeting one Halloween dressed as the Lone Ranger, firing cap pistols to the music of the William Tell Overture, an apt image given that Lewis began to see himself as the Masked Man.[30] Lewis jumps off the pages of our research materials almost like a self-absorbed teenage male who inherits his family's company and turns it into a hedonistic party house in some adolescent, fantasy, B-grade movie.

Yet despite his eccentricities and sometimes outlandish behavior, Lewis dedicated himself to one goal above all others, making Progressive a truly great company.[31] And he built the company to be great *without him*. After Lewis engineered a smooth transition to his successor in 2000, Progressive continued to grow, gaining on its competitors, increasing share value, and sustaining a high return on equity.[32] Did Lewis have a large personal ego and colorful personality? Yes. Did he mature, so that he eventually channeled his ego into building a company that could be great without him? Yes.

The 10Xers share Level 5 leaders' most important trait: they're incredibly ambitious, but their ambition is first and foremost for the cause, for the company, for the work, *not themselves*. Whereas *Good to Great* focused heavily on the humility aspect of Level 5 leaders, this work highlights their sheer ferocity of will.

Sometimes the 10Xers painted their causes in fairly grand terms, even while avoiding any sense of personal grandiosity. Gordon Moore, CEO of Intel from the mid-1970s to mid-1980s, maintained a low profile, despite being the primary company builder during Intel's early growth. Moore nonetheless saw Intel's purpose in gigantic terms, recognizing how microelectronics would revolutionize nearly every aspect of society. In 1973, only five years into Intel's history, Moore said, "We are really the revolutionaries in the world today—not the kids with the long hair and beards who were wrecking the schools a few years ago." Gordon Moore led with an understated personality, yet built a great company that would play a catalytic role in revolutionizing the way civilization works.[33]

To focus on Gordon Moore's understated personality, or Lewis's and Kelleher's outsized personalities, would miss the point. The central question is, *"What are you in it for?"* 10X leaders can be bland or colorful, uncharismatic or magnetic, understated or flamboyant, normal to the point of dull, or just flat-out weird—none of this really matters, as long as they're passionately driven for a cause beyond themselves.

Every 10Xer we studied aimed for much more than just "becoming successful." They didn't define themselves by money. They didn't define themselves by fame. They didn't define themselves by power. They defined themselves by impact and contribution and purpose. Even the über-ambitious Bill Gates, who became the wealthiest person in the world, wasn't driven primarily by gratifying his personal ego. Early in Gates's career, as Microsoft began to gain momentum, one of his friends commented, "All Bill's ego goes into Microsoft. It's his firstborn child."[34] Then later, after working tirelessly for a quarter of a century to make Microsoft a great company, creating powerful software and contributing to the vision of a computer on every desk, he turned with his wife, Melinda, to the question, "How can we do the most good for the greatest number with the resources we have?" And they set forth the audacious aim, among other goals, to eradicate malaria from the face of the Earth.[35]

HOW DO PEOPLE BECOME 10XERS?

We wondered whether the 10Xers had commonalities in their upbringings that might have prepared them for thriving in uncertainty. John Brown of Stryker, for instance, grew up in rural Tennessee, and his family struggled just to have enough food and clothing. "Coming from a modest background makes you focus on the essentials," he later reflected. "I do know what life is like in a ditch [so] I don't get caught up in the fanfare of whether fame and fortune will come." Perhaps someone who rises from a ditch in impoverished rural Tennessee to become a chemical engineer and who then becomes a successful CEO develops an Amundsen-like self-discipline to overcome all odds.[36]

But not every 10Xer grew up in austerity. Herb Kelleher grew up solidly middle class, his father a manager for the ever-stable Campbell Soup Company. He studied philosophy and literature at Wesleyan, graduating with honors as student-body president, and then excelled at NYU Law School, joining the law review and landing a clerkship with the Supreme Court of New Jersey.[37] Peter Lewis grew up in a comfortable home in Cleveland, Ohio, and attended Princeton before taking over the family business.[38]

Furthermore, we found some of the comparison-company leaders had tough early experiences. Yes, John Brown had to climb his way out of a ditch, but his comparison counterpart Leon Hirsch of USSC hardly started from a lofty perch. With a high school education, he'd managed to run only a struggling dry-cleaning-equipment business before he started USSC.[39] Jerry Sanders of comparison case AMD grew up in a gang-infested part of Chicago. In one incident at a party after a football game, Sanders leapt to help a friend who'd gotten himself into a street fight with a gang leader. The friend ran away just as Sanders threw himself into the fray. The gang broke Sanders's nose, fractured his jaw, cracked his skull, cut him up with a beer-can opener, and threw him into a dumpster. Sanders lost so much blood that the hospital called in a priest to read him last rites.[40]

In short, we found no consistent pattern in the backgrounds of 10Xers relative to the comparison leaders. 10Xers can come from tough upbringings or they can come from privileged lives or something in the middle. Nor did we find that they necessarily started as 10Xers; some of the 10Xers evolved, developing their leadership capabilities over time. Herb Kelleher made some terrible decisions early in his career, such as buying Muse Air. Peter Lewis followed a huge arc of growth over three decades and also made some enormously costly blunders along the way. George Rathmann, founder of Amgen, didn't exhibit 10X leadership genius from early on. He'd been denied admission to medical school, so he turned to chemistry as Plan B. He spent 21 years at 3M "highly regarded [but] never considered a star" (according to *Business Week*) and then joined Litton Industries. He floundered in Litton's chaotic culture of acquisitions and later reflected, "I left before I was escorted out."[41]

When we shared the core 10Xer behaviors with our students, former research-team members, and critical readers, we received a series of questions: "Are the 10Xer core behaviors learnable?" "Can anyone become a 10Xer?" "Is it OK to be a 3Xer rather than a 10Xer?" "Do you absolutely need to be a 10Xer to survive a chaotic world?" "Are 10Xers happy?" And so on. We understand these questions, but our research method isn't geared to answer them.

That said, we believe that you do not need answers to these questions to get going. The coming chapters map to the three core 10Xer behaviors, offering practical methods used by these remarkable leaders to build their companies. If your enterprise fully engages these concepts and practices, it'll look a whole lot like a company led by a 10Xer. So, our guidance is simple: get to work learning and applying the practical lessons of how 10Xers lead, building a truly great organization that delivers superior results, makes a distinctive impact, and achieves lasting endurance. There are lots of individually successful people but very few truly great companies that make a 10X impact.

10XERS

KEY POINTS

▶ We named the winning protagonists in our research "10Xers" (pronounced "ten-EX-ers") because they built enterprises that beat their industry's averages by at least 10 times.

▶ The contrast between Amundsen and Scott in their epic race to the South Pole is an ideal analogy for our research question, and a remarkably good illustration of the differences between 10Xers and their comparison companies.

▶ Clear-eyed and stoic, 10Xers accept, without complaint, that they face forces beyond their control, that they cannot accurately predict events, and that nothing is certain; yet they utterly reject the idea that luck, chaos, or any other external factor will determine whether they succeed or fail.

▶ 10Xers display three core behaviors that, in combination, distinguish them from the leaders of the less successful comparison companies:

- *Fanatic discipline:* 10Xers display extreme consistency of action—consistency with values, goals, performance standards, and methods. They are utterly relentless, monomaniacal, unbending in their focus on their quests.

- *Empirical creativity:* When faced with uncertainty, 10Xers do not look primarily to other people, conventional wisdom, authority figures, or peers for direction; they look primarily to empirical evidence. They rely upon direct observation, practical experimentation, and direct engagement with tangible evidence. They make their bold, creative moves from a sound empirical base.

- *Productive paranoia:* 10Xers maintain hypervigilance, staying highly attuned to threats and changes in their environment, even when—especially when—all's going well. They assume conditions will turn against them, at perhaps the worst possible moment. They channel their fear and worry into action, preparing, developing contingency plans, building buffers, and maintaining large margins of safety.

▶ Underlying the three core 10Xer behaviors is a motivating force: passion and ambition for a cause or company larger than themselves. They have egos, but their egos are channeled into their companies and their purposes, not personal aggrandizement.

UNEXPECTED FINDINGS

▶ Fanatic discipline is not the same as regimentation, measurement, obedience to authority, adherence to social stricture, or compliance with bureaucratic rules. True discipline requires mental independence, and an ability to remain consistent in the face of herd instinct and social pressures. Fanatic discipline often means being a nonconformist.

▶ Empirical creativity gives 10Xers a level of confidence that, to outsiders, can look like foolhardy boldness; in fact, empirical validation allows them to simultaneously make bold moves *and* bound their risk. Being empirical doesn't mean being indecisive. 10Xers don't favor analysis over action; they favor empiricism as the foundation for decisive action.

CHAPTER SUMMARY

▶ Productive paranoia enables creative action. By presuming worst-case scenarios and preparing for them, 10Xers minimize the chances that a disruptive event or huge piece of bad luck will stop them from their creative work.

ONE KEY QUESTION

▶ Rank-order the core 10Xer behaviors—fanatic discipline, empirical creativity, and productive paranoia—from your strongest to weakest. What can you do to turn your weakest into your strongest?

3

20 MILE MARCH

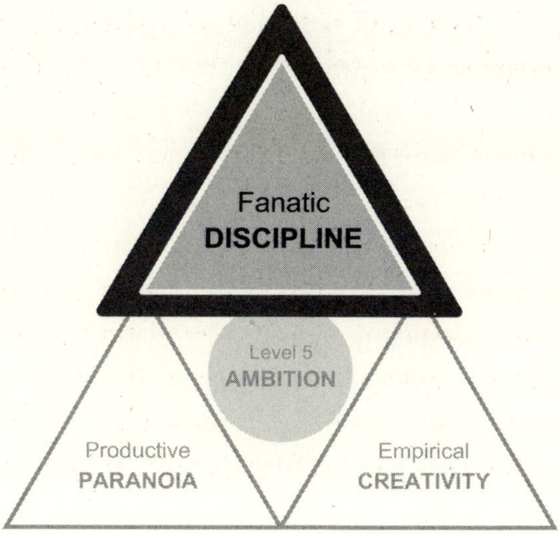

Freely chosen, discipline is absolute freedom.

—Ron Serino[1]

Suppose you have the opportunity to invest in one of two companies, Company A or Company B. Both companies are small, operating in a fast-growing new industry, spinning out disruptive technologies, thriving on rapidly growing customer demand. They have similar product categories, customers, opportunities, and threats; they're a near-perfect matched pair.

Company A will achieve 25 percent average annual growth in net income over a 19-year period.

Company B will achieve 45 percent average annual growth in net income over the same 19 years.

Stop and think: which company will you want to invest in?

Most people, including us, would invest in Company B, given no additional information.

Now, let's add some more information.

The standard deviation of net income growth (which reflects the degree of volatility) for Company A over that period will be 15 percentage points.

The standard deviation for Company B over the same years will be 116 percentage points.

Company A will maintain consistent and controlled growth, staying *below* 30 percent for 16 of 19 years yet achieving 20 percent or more almost every year. Company B will show a much more erratic and uncontrolled growth pattern than Company A. Company B's annual net income growth rate will *exceed* 30 percent for 13 of 19 years, with net income growth rates ranging from positive 313 percent to negative 200 percent.[2]

By now, you're probably suspecting that Company A turned out to be a better investment than Company B, despite the fact that B generally grew faster. And you'd be correct. But the amazing thing is *how much* better. Look at the chart "Value of $1 Invested, Company A vs. Company B."

Company A is Stryker and Company B is USSC. Every $1 invested in Stryker at the end of 1979 (the year of its initial public offering) and held through 2002 multiplied more than 350 times. Every $1 invested in USSC on the same date generated cumulative returns that fell below the general market by 1998, and then . . . it disappeared from the chart. For all its extraordinary growth, USSC capitulated to an acquisition, giving up forever its chance to come back as a great company.[3]

Value of $1 Invested
Company A vs. Company B

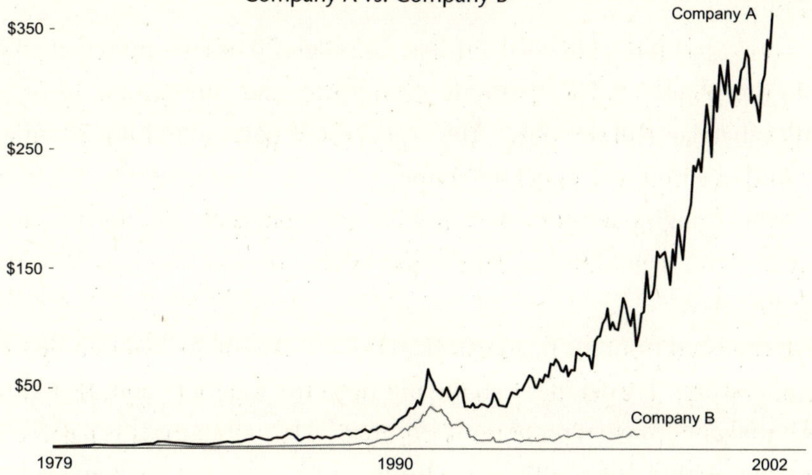

Notes:
1. Source for all stock return calculations in this work: ©200601 CRSP®, Center for Research in Security Prices. Booth School of Business, The University of Chicago. Used with permission. All rights reserved. www.crsp.chicagobooth.edu.

JOHN BROWN'S 20 MILE MARCH

Imagine you're standing with your feet in the Pacific Ocean in San Diego, California, looking inland. You're about to embark on a three-thousand-mile walk, from San Diego to the tip of Maine.

On the first day, you march 20 miles, making it out of town.

On the second day, you march 20 miles. And again, on the third day, you march 20 miles, heading into the heat of the desert. It's hot, more than a hundred degrees, and you want to rest in the cool of your tent. But you don't. You get up and you march 20 miles.

You keep the pace, 20 miles a day.

Then the weather cools, and you're in comfortable conditions with the wind at your back, and you could go much farther. But you hold back, modulating your effort. You stick with your 20 miles.

Then you reach the Colorado high mountains and get hit by snow, wind, and temperatures below zero—and all you want to do is stay in your tent. But you get up. You get dressed. You march your 20 miles.

You keep up the effort—20 miles, 20 miles, 20 miles—then you cross into the plains, and it's glorious springtime, and you can go 40 or 50 miles in a day. But you don't. You sustain your pace, marching 20 miles.

And eventually, you get to Maine.

Now, imagine another person who starts out with you on the same day in San Diego. He gets all excited by the journey and logs 40 miles the first day.

Exhausted from his first gigantic day, he wakes up to hundred-degree temperatures. He decides to hang out until the weather cools, thinking, "I'll make it up when conditions improve." He maintains this pattern—big days with good conditions, whining and waiting in his tent on bad days—as he moves across the western United States.

Just before the Colorado high mountains, he gets a spate of great weather and he goes all out, logging 40- to 50-mile days to make up lost ground. But then he hits a huge winter storm when utterly exhausted. It nearly kills him and he hunkers down in his tent, waiting for spring.

When spring finally comes, he emerges, weakened, and stumbles off toward Maine. By the time he enters Kansas City, you, with your relentless 20 mile march, have already reached the tip of Maine. You win, by a huge margin.

Now, think of Stryker as a *20 Mile March* company.

When John Brown became CEO of Stryker in 1977, he deliberately set a performance benchmark to drive consistent progress: Stryker would achieve 20 percent net income growth every year. This was more than a mere target, or a wish, or a hope, or a dream, or a vision. It was, to use Brown's own words, "the law." He ingrained "the law" into the company's culture, making it a way of life.[4]

Brown created the "Snorkel Award," given to those who lagged behind; 20 percent was the watermark, and if you were below the watermark, you needed a snorkel. Just imagine receiving a mounted snorkel from John Brown to hang on your wall so everyone can see that you're in danger of drowning. People worked hard to keep the snorkel off their walls.[5]

Imagine going to a big company meeting. You walk into the main ballroom to find sales regions arranged by performance. Those in regions that achieved their 20 Mile March get seating assignments at the front of the room; those that fell behind find themselves assigned to tables in the back of the room.[6]

Stryker's annual division-review meetings included a chairman's breakfast. Those who hit their 20 Mile March went to John Brown's breakfast table. Those who didn't went to another breakfast. "They are well fed," said Brown, "but it is not the one where you want to go."[7]

If your division fell behind for two years in a row, Brown would insert himself to "help," working around the clock to "help" you get back on track. "We'll arrive at an agreement as to what has to be done to correct the problem," said the understated Brown. You get the distinct impression that you really don't want to need John Brown's help. According to *Investor's Business Daily*, "John Brown doesn't want to hear excuses. Markets bad? Currency exchange rates are hurting results? Doesn't matter." Describing challenges Stryker faced in Europe due partly to currency exchange rates, an analyst noted, "It's hard to know how much of [the problem] was external. But at Stryker, that's irrelevant."[8]

From the time John Brown became CEO in 1977 through 1998 (when its comparison, USSC, disappeared as a public company) and excluding a 1990 extraordinary gain, Stryker hit its 20 Mile March goal more than 90 percent of the time. Yet for all this self-imposed pressure, Stryker had an equally important self-imposed constraint: to never go too far, to never grow too much in a single year. Just imagine the pressure from Wall Street to increase growth when your direct rival is growing faster than your company. In fact, Stryker grew *more slowly* than USSC more than half the time. According to the *Wall Street Transcript*, some observers criticized Brown for not being more aggressive. Brown, however, consciously chose to maintain the 20 Mile March, regardless of criticism urging him to grow Stryker at a faster pace in boom years.[9]

John Brown understood that if you want to achieve consistent performance, you need both parts of a 20 Mile March: a lower bound *and* an upper bound, a hurdle that you jump over *and* a ceiling that you will not rise above, the ambition to achieve *and* the self-control to hold back.

It would be hard to find a more perfect, stark contrast to Stryker than the spectacular rise and fall of USSC. In 1989, USSC had $345 million in sales; in 1992, it had $1.2 billion, representing 248 percent growth in just three years. USSC aggressively pursued growth, betting on a new line of sutures in a direct attack on Johnson & Johnson's Ethicon division, which controlled 80 percent of the sutures business. At the time, a toehold of even 10 percent of the market would have added 40 percent to USSC sales, but USSC's founder, Leon Hirsch, scoffed at such small thinking, "I'd be disappointed if we got just 10% [of the sutures market]—and Ethicon would be elated." USSC pushed inventories onto hospitals, so much so that the *Wall Street Journal* reported, "According to the lore surrounding USSC's reputation for aggressive marketing, a salesman aiming to boost volume once hid so much inventory in a hospital storeroom's false ceiling that it collapsed." The company also attained explosive growth from the rapid adoption of its laparoscopic instruments for gallbladder surgery (laparoscopy is a minimally invasive surgical technique), and it sought even more growth by expanding the use of its laparoscopic instruments into a range of other surgical procedures.[10]

But then—bang!—USSC got walloped by a series of storms. The specter of the Clinton healthcare reform created uncertainty, and hospitals decreased purchasing. Doctors showed less-than-expected enthusiasm for new laparoscopic devices for other than gallbladder surgery. Johnson & Johnson proved to be a formidable competitor in sutures, striking back hard, holding on to much of its market share. Johnson & Johnson also attacked USSC's core laparoscopic business, taking 45 percent of domestic market share in just three years. Revenues fell, and by

1997, they remained below peak 1992 levels. By the end of 1998, USSC would no longer exist as an independent company, acquired by Tyco.[11]

20 MILE MARCH—NOT WHAT WE EXPECTED

When we began this study, we thought we might see 10X winners respond to a volatile, fast-changing world full of new opportunities by pursuing aggressive growth and making radical, big leaps, catching and riding the Next Big Wave, time and again. And yes, they did grow, and they did pursue spectacular opportunities as they grew. But the less successful comparison cases pursued *much more* aggressive growth and undertook big-leap, radical-change adventures to a much greater degree than the 10X winners. The 10X cases exemplified what we came to call the 20 Mile March concept, hitting stepwise performance markers with great consistency over a long period of time, and the comparison cases did not.

> The 20 Mile March is more than a philosophy. It's about having concrete, clear, intelligent, and rigorously pursued performance mechanisms that keep you on track. The 20 Mile March creates two types of self-imposed discomfort: (1) the discomfort of unwavering commitment to high performance in difficult conditions, and (2) the discomfort of holding back in good conditions.

Southwest Airlines, for example, demanded of itself a profit every year, even when the entire industry lost money. From 1990 through 2003, the U.S. airline industry as a whole turned a profit in just 6 of 14 years. In the early 1990s, the airline industry lost $13 billion and furloughed more than a hundred thousand employees; Southwest remained profitable and furloughed not a single person. Despite an almost chronic epidemic of airline troubles, including high-profile bankruptcies of some major carriers, Southwest generated a profit every year for 30 consecutive years.[12]

Equally important, Southwest had the discipline to hold back in good times so as not to extend beyond its ability to preserve profitability and the Southwest culture. It didn't expand outside Texas until nearly eight years after starting service, making a small jump to New Orleans. Southwest moved outward from Texas in deliberate steps—Oklahoma City, Tulsa, Albuquerque, Phoenix, Los Angeles—and didn't reach the eastern seaboard until almost a quarter of a century after its founding. In 1996, more than a *hundred* cities clamored for Southwest service. And how many cities did Southwest open that year? Four. (See diagram "Southwest Airlines 20 Mile March.")[13]

At first glance, this might not strike you as particularly significant. But stop to think about it. Here we have an airline setting for itself a standard of consistent performance that no other airline achieves. Anyone who said they'd be profitable *every* year for nearly three decades in the airline business—the *airline* business!—would be laughed at. No one does that. But Southwest did. Here also we have a publicly traded company willing to leave growth on the table. How many business leaders of publicly traded companies have the ability to leave gobs of growth on the table, especially during boom times when competitors do *not* leave growth on the table? Few, indeed. But Southwest did that, too.[14]

> Some people believe that a world characterized by radical change and disruptive forces no longer favors those who engage in consistent 20 Mile Marching. Yet the great irony is that when we examined just this type of out-of-control, fast-paced environment, we found that every 10X company exemplified the 20 Mile March principle during the era we studied.

Now, you might be wondering, "But wait a minute! You're confusing things here. Perhaps 10X companies could afford to behave this way because they were so successful and dominant. Perhaps 20 Mile Marching is a result of success, a luxury of success, not a driver of success." But the evidence shows the 10X companies embraced a 20 Mile March early, long *before* they were big companies.

Furthermore, every comparison company failed to 20 Mile March with anything close to the consistency shown by the 10X cases. In fact, this is one of the strongest contrasts in our study. (See *Research Founda-*

Southwest Airlines 20 Mile March

Part 1: Stepwise Performance in Good Times and Bad – Profitable Every Year

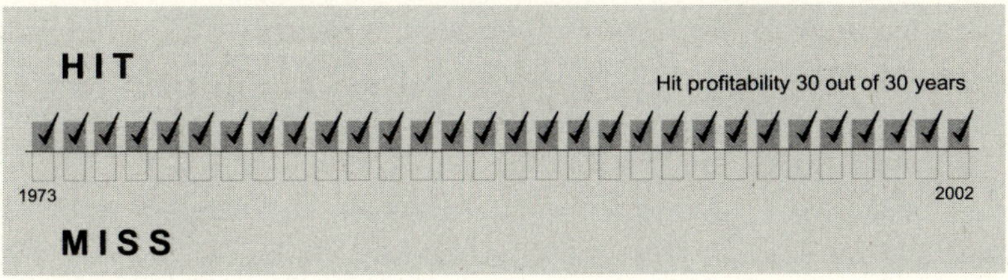

Southwest Airlines 20 Mile March

Part 2: The Discipline to Hold Back – Did Not Expand Too Much in a Given Year

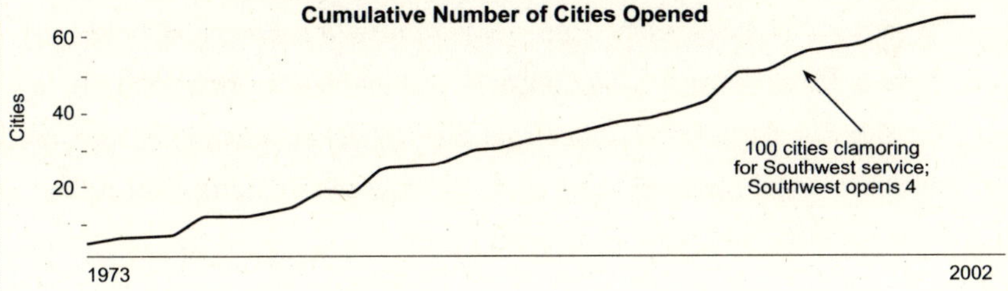

Southwest Airlines 20 Mile March Outcome

Part 3: Cumulative Stock Returns

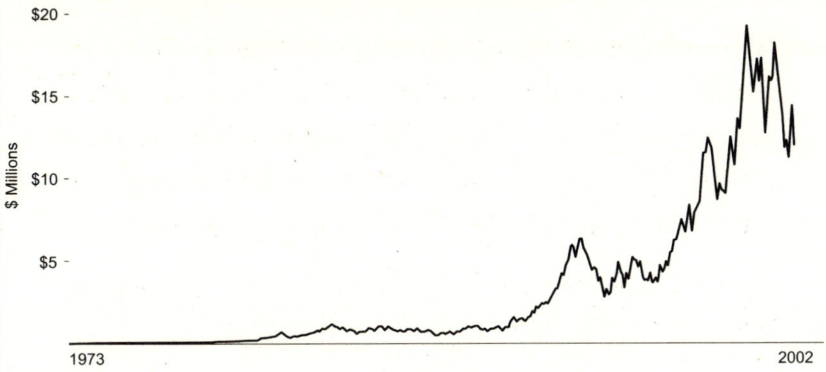

Notes:

1. Value of $10,000 invested, December 31, 1972 through December 31, 2002.
2. Source for all stock return calculations in this work: ©200601 CRSP®, Center for Research in Security Prices. Booth School of Business, The University of Chicago. Used with permission. All rights reserved. www.crsp.chicagobooth.edu.

tions: 20 Mile March Analysis.) Some comparisons showed no sign of 20 Mile Marching at any time during the era of study, as with USSC, AMD, and Kirschner. Some comparisons showed no 20 Mile Marching during their worst years, only to regain ground when they finally became 20 Mile Marchers, as with Genentech under Arthur Levinson and Apple under Steve Jobs. Other comparison companies, such as PSA and Safeco, 20 Mile Marched in their early years, when they produced their best results, then later fell behind when they lost discipline.

ELEMENTS OF A GOOD 20 MILE MARCH

A good 20 Mile March uses *performance markers* that delineate a lower bound of acceptable achievement. These create productive discomfort, much like hard physical training or rigorous mental development, and must be challenging (but not impossible) to achieve in difficult times.

A good 20 Mile March has *self-imposed constraints.* This creates an upper bound for how far you'll march when facing robust opportunity and exceptionally good conditions. These constraints should also produce discomfort in the face of pressures and fears that you should be going faster and doing more.

A good 20 Mile March is *tailored to the enterprise* and its environment. There's no all-purpose 20 Mile March for all enterprises. Southwest's march wouldn't apply to Intel. A sports team's march wouldn't apply to an Army platoon leader. An Army platoon leader's march wouldn't apply to a school.

A good 20 Mile March *lies largely within your control to achieve.* You shouldn't need luck to achieve your march.

A good 20 Mile March has *a Goldilocks time frame,* not too short and not too long but just right. Make the timeline of the march too short, and you'll be more exposed to uncontrollable variability; make the timeline too long, and it loses power.

A good 20 Mile March is *designed and self-imposed by the enterprise,* not imposed from the outside or blindly copied from others. For instance, to simply accept "earnings per share" as the focus of a march because Wall Street looks at earnings per share would lack rigor, reflecting no clarity about the underlying performance drivers in a specific enterprise.

A good 20 Mile March must be *achieved with great consistency.* Good intentions do not count.

WHAT MAKES A GOOD 20 MILE MARCH

In the early 1970s, Peter Lewis articulated a stringent performance metric: Progressive Insurance should grow only at a rate at which it could still sustain exemplary customer service and achieve a profitable "combined ratio" averaging 96 percent. What does a combined ratio of 96 percent mean? If you sell $100 of insurance, you should need to pay out no more than $96 in losses plus overhead combined. The combined ratio captures the central challenge for the insurance business, pricing premiums at a rate that'll allow you to pay out on losses, service customers, and earn a return. If a company lowers prices to increase growth, its combined ratio could deteriorate. If it misjudges risks or mismanages its claims service, its combined ratio will suffer. If the combined ratio climbs over 100 percent, the company loses money on its insurance business.[15]

Progressive's "profitable combined ratio" mantra became like John Brown's 20 percent law, a rigorous standard to accomplish year in and year out. Progressive's stance: If competitors lower rates in an unprofitable bid to increase share—fine, let them do so! We will not chase them into senseless self-destruction. Progressive had an unequivocal commitment to the profitable combined ratio, no matter what conditions it faced, how its competitors behaved, or what seductive growth opportunities beckoned. Said Lewis in 1972, "There is no excuse, not regulatory problems, not competitive difficulties, not natural disaster, for failing to do so." Progressive achieved a profitable combined ratio 27 out of 30 years 1972–2002, and averaged just better than its 96 percent target.[16]

Now, compare Progressive's combined-ratio discipline to the criteria outlined in the table "Elements of a Good 20 Mile March."

Performance marker: check.
Self-imposed constraint: check.
Appropriate to the enterprise: check.
Largely within its own control: check.
Goldilocks time frame: check.
Designed and self-imposed by the enterprise: check.
Achieved with high consistency: check.

A 20 Mile March operates as a practical, powerful strategic mechanism. John Brown built the entire Stryker system, from rapid product-development cycles to the Snorkel Award, to achieve "the law" (20 percent earnings growth). Peter Lewis designed his entire system so as to achieve the 96 percent combined ratio. "It sounds simple, but it is very difficult to do," said Lewis's successor, Glenn M. Renwick. "Think about it as a recipe. If you over-weight any of the ingredients, you won't get the result you wanted. Think what a disaster it would be to realize you messed up only one ingredient, but it was four times as much as you should have put in . . . A 96% combined ratio means we have to be disciplined in every segment of our business. It means we say that we'd rather be consistently growing . . . than be hot for one year and then gone the next."[17]

Do you need to accomplish your 20 Mile March with 100 percent success? The 10X companies didn't have a perfect record, only a near-perfect record, but they never saw missing a march as "OK." If they missed it even once, they obsessed over what they needed to do to get back on track: *There's no excuse, and it's up to us to correct for our failures, period.*

Consider the sad demise of Progressive's comparison case, Safeco. Prior to the 1980s, Safeco displayed a "nearly fanatic" dedication to

The 20 Mile March imposes order amidst disorder, consistency amidst swirling inconsistency. But it works only if you actually achieve your march year after year. If you set a 20 Mile March and then fail to achieve it—or worse, abandon fanatic discipline altogether—you may well get crushed by events.

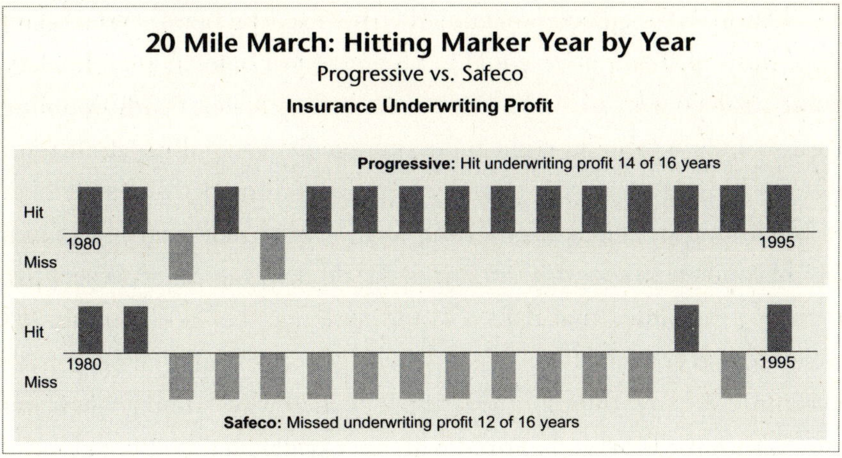

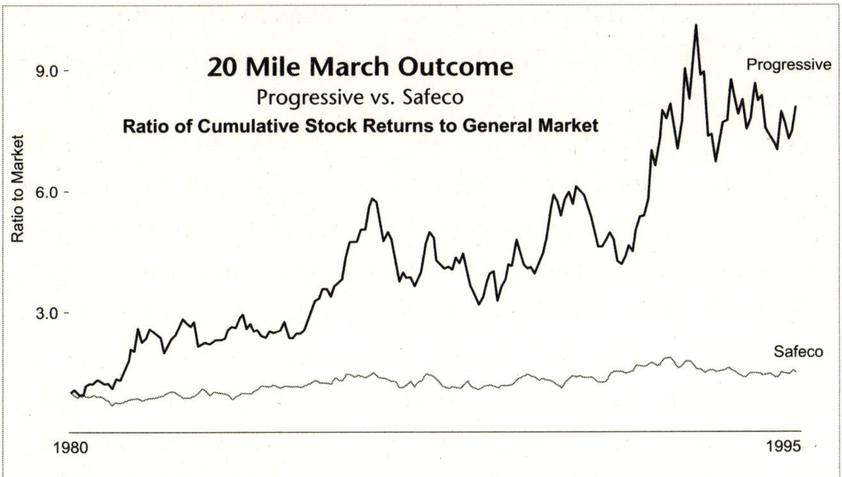

Notes:

1. Each company's ratio to market was calculated from December 31, 1979 to December 31, 1995.
2. Source for all stock return calculations in this work: ©200601 CRSP®, Center for Research in Security Prices. Booth School of Business, The University of Chicago. Used with permission. All rights reserved. www.crsp.chicagobooth.edu.

a profitable combined ratio, in good times and bad, very much like Progressive. Then in the 1980s, Safeco lost its discipline. (The diagram "20 Mile March: Hitting Marker Year by Year" shows the point of divergence in the 1980s and early 1990s, when Safeco lost discipline while Progressive did not.) It failed to consistently achieve its combined ratio, became seduced by spectacular returns from investing insurance premiums in the capital markets and fell behind in its core business. In 1989, for instance, Safeco lost $52 million on its core underwriting business yet made $263 million in profits from its investment portfolio.[18]

Then in 1997, Safeco proclaimed "truly exciting news" and a "giant step" forward. For a price equal to 68 percent of Safeco's shareholders' equity, Safeco won an auction to buy American States, nearly doubling its distribution force to eight thousand agents, catapulting itself from #22 to #12 in property and casualty insurance, jumping from a regional to a national presence, and setting forth a bold new goal of expanding beyond insurance into financial products. One Safeco executive proudly proclaimed that Safeco would no longer be "dull, boring, traditional, and conservative." After all, why get back to all that pedestrian discipline, why go through the struggle of a 20 Mile March, when you can make up all that lost ground with one spectacular and imaginative jump? Heralding the great leap forward, Safeco's CEO Roger Eigsti opened his 1997 annual letter to shareholders, "Future generations will chronicle 1997 as a remarkable year for Safeco."[19]

It was indeed a remarkable turning point, but not the one Eigsti envisioned. The combined ratio suffered, unprofitable in 1998, 1999, 2000, 2001, and 2002. "We perhaps pushed too hard for growth," said one executive of Safeco's slide. Every dollar invested in Safeco at the start of 1997, the year of the American States acquisition, lost 30 percent of its value over the next three years, falling more than 60 percent behind the general stock market. Three years after Safeco's big, bold leap, Eigsti announced his retirement and the board launched a search for a new CEO, eventually going outside for a savior to turn the company around. From the beginning of 1976 through 2002, Safeco achieved a profitable combined ratio in only 10 of 27 years; over those same years, Progressive—the boring, consistent-combined-ratio champion—generated cumulative returns to investors nearly 32 times greater than Safeco.[20]

While the 20 Mile Marches we've discussed so far—Stryker's 20 percent earnings growth, Southwest's profit every year, and Progressive's 96 percent combined ratio—are financial, we want to be clear that you can also have a non-financial march. A school might have a student-performance march. A hospital might have a patient-safety march. A church might have a number-of-converts march. A government agency might have a continuous-improvement march. A homeless center might have a getting-people-housed march. A police department might have a crime-rate march. Corporations, too, can choose a non-financial march, such as an innovation march. Intel, for instance, built its 20 Mile March around the idea of "Moore's Law" (double the complexity of components per integrated circuit at an affordable cost every 18 months to two years). Intel sustained its commitment to achieving Moore's Law whether in boom times or industry depression, retaining its best engineers, always moving to the next-generation chip, investing consistently in its creative march, year in and year out, no matter what, for more than thirty years.[21]

20 MILE MARCH CONTRASTS THROUGH 2002

10X Case	Comparison Case
<u>Stryker</u> Achieved 20% annual earnings' growth. Also practiced 20 Mile March innovation via lots of product iterations and extensions. Held back on growth in good times, which enabled it to weather difficult industry events from 1992 to 1994.[22]	<u>USSC</u> Experienced erratic earnings' growth. Sought big breakthrough innovation rather than 20 Mile March innovation. Overextended in difficult times, especially from 1992 to 1994; sold out in 1998.[23]
<u>Southwest Airlines</u> Achieved profitability for 30 consecutive years. Unlike the other major airlines, turned a profit in 2002 in the aftermath of 9/11. Constrained growth to ensure profitability and preserve culture.[24]	<u>PSA</u> Had a 20 Mile March philosophy with consistent profitability in its early history but abandoned it in the 1970s. Capitulated to a takeover by US Air in 1986.[25]

(*continued on next page*)

10X Case	Comparison Case
Progressive Insurance Kept combined ratio below 100% every year, averaging 96% across time. Achieved profitable combined ratio in 27 out of 30 years. Limited growth to ensure that it maintained underwriting standards and hit combined-ratio objective.[26]	*Safeco Insurance* Focused on combined ratio in its early history. From 1980 on, became inconsistent, then went for big growth via huge acquisition of American States in the 1990s. Attained profitable combined ratio in only 10 of 27 years.[27]
Intel Upheld Moore's Law, doubling the complexity of components per integrated circuit at minimum cost every 18 months to two years. Pursued this relentlessly over the entire era of our analysis.[28]	*AMD* Repeatedly pursued big growth in good times (sometimes with significant debt), leaving company unprepared for bad times (especially 1985–1986). No evidence of steady performance marker.[29]
Microsoft Practiced 20 Mile March innovation, consisting of continuous iterations of software products. Often began with imperfect products, then marched to improve year after year to achieve eventual industry dominance. Never overextended financially, so never needed to pause its march.[30]	*Apple* Didn't 20 Mile March during its early history. Experienced inconsistent profit growth, and setbacks in the mid-1980s, early-1990s, and mid-1990s. Adopted 20 Mile March innovation with return of Steve Jobs, a key factor in its resurgence in the 2000s.[31]
Amgen Undertook 20 Mile March innovation based on incremental product innovation and product-development milestones. Continuously developed existing drugs for new indications. Resulted in strong revenue growth.[32]	*Genentech* Didn't 20 Mile March from 1976 to 1995, following a bet-big mentality coupled with overpromises, resulting in a downfall. After 1995, followed a 20 Mile March strategy of breaking five-year goals into a series of one-year targets.[33]
Biomet Focused on consistent profitable growth, achieved in 20 of 21 years. Also practiced 20 Mile March innovation, with rapid product-development iterations. Took care never to overextend.[34]	*Kirschner* Didn't 20 Mile March. Embarked on a "grow fast through acquisition" approach, using debt. Resulted in crisis and sale of the company in 1994.[35]

WHY 20 MILE MARCHERS WIN

20 Mile Marching helps turn the odds in your favor for three reasons:

1. It builds confidence in your ability to perform well in adverse circumstances.
2. It reduces the likelihood of catastrophe when you're hit by turbulent disruption.
3. It helps you exert self-control in an out-of-control environment.

CONFIDENCE BUILT FROM PERFORMANCE IN ADVERSITY

Confidence comes not from motivational speeches, charismatic inspiration, wild pep rallies, unfounded optimism, or blind hope. Taciturn, understated, and reserved, John Brown at Stryker avoided all of these. Stryker earned its confidence by actual achievement, accomplishing stringent performance standards year in and year out, no matter the industry conditions. John Brown operated like a track coach who trains his runners to run strong at the end of every workout, in wind, in heat, in rain, in snow, no matter what the conditions. And then if it's windy, hot, rainy, or snowy on championship day, the runners feel confident because of their own actual experience: we *can* run strong because we've trained hard even when we felt bad, because we've practiced running hard in heinous conditions!

> Accomplishing a 20 Mile March, consistently, in good times and bad, builds confidence. Tangible achievement in the face of adversity reinforces the 10X perspective: *we are ultimately responsible* for improving performance. We never blame circumstance; we never blame the environment.

In 2002, we received a phone call at our research lab in Boulder, Colorado, from Lattie Coor, former president of Arizona State University and then chairman of the Center for the Future of Arizona. "We've identified the education of Latino children as one of our state's top priorities," said Coor. "We must figure out how to solve the problem. Can you give us some guidance?" Coor had the idea to create a study patterned on a matched-pair method similar to this study's but applied to education. They'd identify public schools that performed well in adverse circumstances and with significant Latino populations; they'd then compare those schools to other public schools facing similar circumstances that didn't perform as well and study the differences.

Coor assembled a team of researchers led by Mary Jo Waits, who conducted its *Beat the Odds* study with guidance from our research lab.[36] The study found that factors outside principals' control—such as class size, the length of school day, the amount of funding, and the degree of parental involvement—did *not* systematically distinguish the higher performing from the comparison schools. Of course, changing those variables might well improve education performance across all schools, but the beat-the-odds schools put their energies into what *they could do*. The study identified a set of practical disciplines that lay within the control of the individual school, *even* in adverse circumstances. Each beat-the-odds school held itself accountable for a clear bottom line of academic performance, rooted in three precepts articulated in the *Beat the Odds* report:

- Don't even think about playing a blame game when students aren't learning. Have the strength to look at the problem and take responsibility.
- Don't think the solution is "out there." If students aren't learning, the school needs to change.
- No one is allowed to lag behind. If every student in every classroom isn't learning, the school isn't doing its job.

In 1997, Alice Byrne Elementary School in Yuma, Arizona, performed no better than a similar comparison school and substantially below state averages in third-grade reading. Principal Juli Tate Peach refused to capitulate to difficult circumstances. Yes, many of the kids came from poor Latino families. Yes, the school had a limited budget. Yes, the teachers felt stretched to do more with less. Peach and her teachers nonetheless overcame these obstacles and gradually increased student reading performance about 20 percentage points, to beat the state averages. Meanwhile, Alice Byrne's comparison school, *facing similar circumstances*, demonstrated no substantial improvement in third-grade reading. Why?

Juli Tate Peach brought fanatic discipline to one focused goal: individual student achievement in basic skills like reading. She led the school to measure progress not just at the end of the year but also throughout the year, working with her teachers to track performance, taking corrective action along the way. She created a collaborative culture of teachers and administrators poring over the data and sharing ideas for how to help each child perform better. They embraced a never-ending cycle of instruction, assessment, intervention, kid by kid, in a relentless 20 Mile March of learning throughout the year. Improving results increased confidence and motivation, which then reinforced discipline, which then drove better results, which then increased confidence and motivation, which then reinforced discipline, up and up and up.

The principals at the Arizona beat-the-odds schools understood that grasping for the next "silver bullet" reform—lurching from one program to the next, this year's fad to next year's fad—destroys motivation and erodes confidence. The critical step lay not in finding the perfect program or in waiting for national education reform, but in taking action; picking a good program; instilling the fanatic discipline to make relentless, iterative progress; and staying with the program long enough to generate sustained results. They gained confidence by the very *fact* of increasing achievement. If you beat the odds, you then gain confidence that you can beat the odds again, which then builds confidence that you can beat the odds again, and again, and again.[37]

AVOIDANCE OF CATASTROPHE

In the 1980s, AMD nearly destroyed itself by failing to 20 Mile March. In 1984, Jerry Sanders proclaimed that AMD would become the first semiconductor company to generate 60 percent growth two years in a row and that it could grow more in a single year than it had in its entire 14-year prior history. Not only that, he announced that AMD would aim to become #1 in integrated circuits by the end of the decade, ahead of Intel, ahead of Texas Instruments, ahead of National Semiconductor, ahead of Motorola, ahead of every American competitor. It was quite a contrast to Intel, where Gordon Moore stated at the exact same time that he aimed to *limit* Intel's growth so as to minimize the chances of losing control. Intel still grew at a rapid rate but held growth back relative to AMD; from 1981 through 1984, AMD grew at nearly twice the rate of Intel and faster than every other American competitor.[38]

Then in 1985, the semiconductor industry collapsed into a recession. Both Intel and AMD suffered, but AMD suffered much worse. Sales fell from $1.1 billion to $795 million within one year.[39] AMD, which had tripled its long-term debt, didn't recover for years. When AMD and Intel emerged from the storm, Intel pulled ahead for good. In the 12 years prior to the 1985 industry meltdown, AMD's stock returns outpaced Intel's, fueled in part by AMD's tripling sales from 1981 through 1984. But coming out of the industry meltdown, AMD fell behind while Intel soared; from the start of 1987 through 1994, Intel's shareholder returns outpaced AMD's by more than five times, then continued on pace to beat AMD by more than thirty times through 2002. (See diagram "Intel's 20 Mile March vs. AMD's Boom and Bust.")[40]

If you deplete your resources, run yourself to exhaustion, and then get caught at the wrong moment by an external shock, you can be in serious trouble. By sticking with your 20 Mile March, you reduce the chances of getting crippled by a big, unexpected shock. Every 10X winner pulled further ahead of its less successful comparison company during turbulent times. Ferocious instability favors the 20 Mile Marchers. This is when they really shine.

Intel's 20 Mile March vs. AMD's Boom and Bust

Ratio of Cumulative Stock Returns to General Market

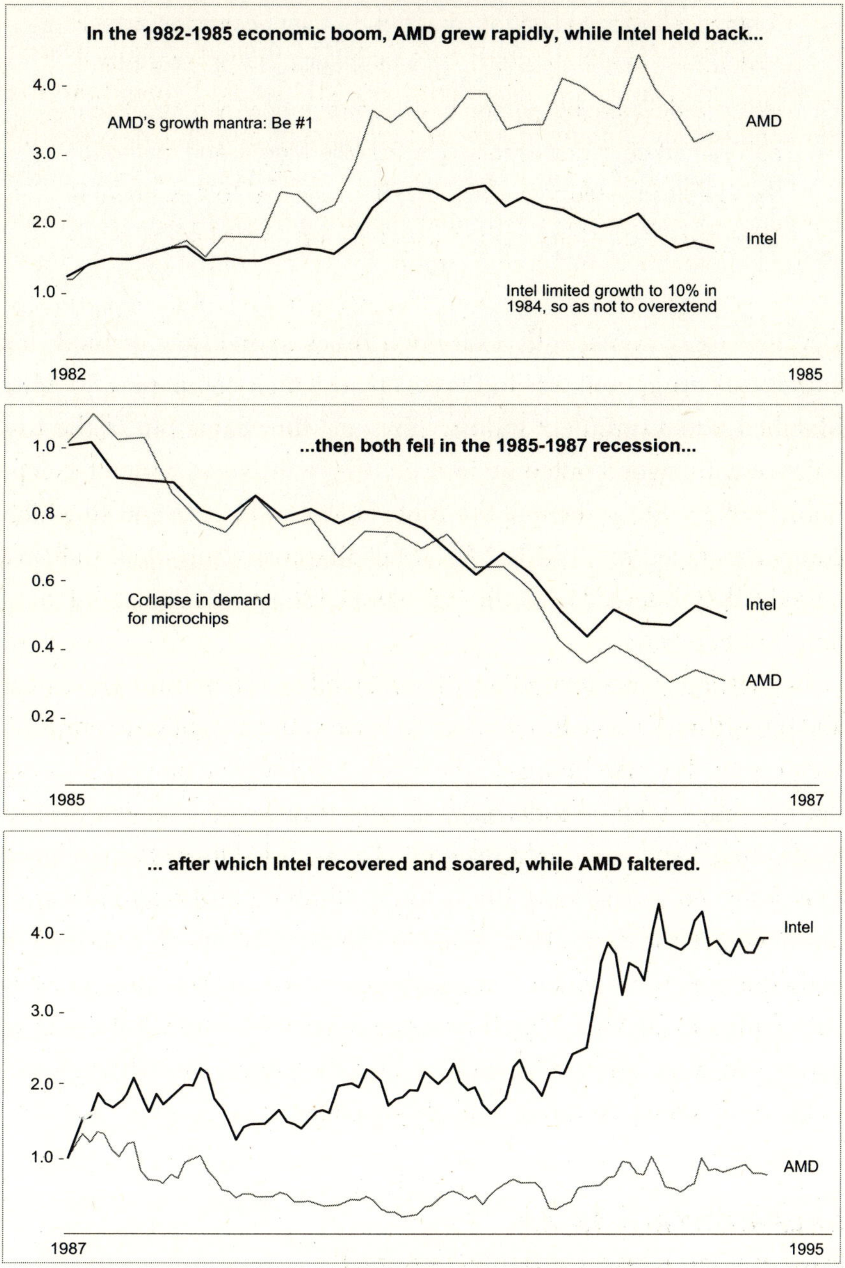

Notes:

1. Chart 1: December 31, 1981 to December 31, 1984. Chart 2: December 31, 1984 to December 31, 1986. Chart 3: December 31, 1986 to December 31, 1994.
2. Source for all stock return calculations in this work: ©200601 CRSP®, Center for Research in Security Prices. Booth School of Business, The University of Chicago. Used with permission. All rights reserved. www.crsp.chicagobooth.edu.

Failure to 20 Mile March in an uncertain and unforgiving environment can set you up for catastrophe. Every comparison case had an episode in its history in which failing to 20 Mile March led to a devastating outcome. In contrast, only two 10X companies had episodes of failing to 20 Mile March, and neither of these episodes led to catastrophe because the 10X companies self-corrected before a storm could rise up and kill them.

When we systematically examined times of industry turmoil, we found a sobering contrast. In 29 events in which companies 20 Mile Marched into a turbulent industry episode, they came out of the turbulence with a good outcome in every single instance, without exception, 29 of 29, 100 percent of the time. However, in 23 events in which companies failed to 20 Mile March heading into a turbulent industry episode, they emerged from the turbulent episode with a good outcome only 3 out of 23 times.

In a setting characterized by unpredictability, full of immense threat and opportunity, you cannot afford to leave yourself exposed to unforeseen events. If you're hiking in the warm, comfortable glow of a spring day on a nice, wide, wandering trail near your home, you can overextend yourself and you might need to take two Advil to soothe your sore muscles when you're done. But if you're climbing in the Himalayas or journeying to the South Pole, going too far can have much more severe consequences from which you might never recover. You can get away with failing to 20 Mile March in stable times for a while, but doing so leaves you weak and undisciplined, and therefore exposed when unstable times come. And they will always come.

SELF-CONTROL IN AN OUT-OF-CONTROL ENVIRONMENT

On December 12, 1911, Amundsen and his team reached a point 45 miles from the South Pole. He had no idea of Scott's whereabouts.

Scott had taken a different route slightly to the west, so for all Amundsen knew, Scott was ahead of him. The weather had turned clear and calm, and sitting high on the smooth Polar Plateau, Amundsen had perfect ski and sled conditions for the remainder of the journey to the South Pole. Amundsen noted, "Going and surface as good as ever. Weather splendid—calm with sunshine." His team had journeyed more than 650 miles, carving a path straight over a mountain range, climbing from sea level to over ten thousand feet. And now, with the anxiety of "Where's Scott?" gnawing away, his team could reach its goal within 24 hours in one hard push.

And what did Amundsen do?

He went 17 miles.

Throughout the journey, Amundsen adhered to a regimen of consistent progress, never going too far in good weather, careful to stay far away from the red line of exhaustion that could leave his team exposed, yet pressing ahead in nasty weather to stay on pace. Amundsen throttled back his well-tuned team to travel between 15 and 20 miles per day, in a relentless march to 90 degrees South. When a member of Amundsen's team suggested they could go faster, up to 25 miles a day, Amundsen said no. They needed to rest and sleep so as to continually replenish their energy. We'd uncovered the 20 Mile March concept in our study fully three years before we stumbled across the Amundsen and Scott story, and we'd been using the term "20 Mile March" in our research discussions, and with clients and students. So, we were astounded to learn that Amundsen had embraced this precise idea in his journey to the South Pole.

In contrast, Scott would sometimes drive his team to exhaustion on good days and then sit in his tent and complain about the weather on bad days. In early December, Scott wrote in his journal about being stopped by a blizzard, "I doubt if any party could travel in such weather." But when Amundsen faced conditions comparable to Scott's (even colder and at higher altitude as he moved through the mountain passes) he wrote in *his* journal, "It has been an unpleasant day—storm, drift and frostbite, but we have advanced 13 miles closer to our goal."

According to Roland Huntford's account in *The Last Place on Earth*, Scott faced 6 days of gale-force winds and traveled on none, whereas Amundsen faced 15 and traveled on 8. Amundsen clocked in at the South Pole right on pace, having averaged 15.5 miles per day.[41]

> Like Amundsen and his team, the 10Xers and their companies use their 20 Mile Marches as a way to exert self-control, even when afraid or tempted by opportunity. Having a clear 20 Mile March focuses the mind; because everyone on the team knows the markers and their importance, they can stay on track.

Financial markets are out of your control. Customers are out of your control. Earthquakes are out of your control. Global competition is out of your control. Technological change is out of your control. Most everything is ultimately out of your control. But when you 20 Mile March, you have a tangible point of focus that keeps you and your team moving forward, despite confusion, uncertainty, and even chaos.

ARTHUR LEVINSON: TEACHING A COMPANY TO MARCH

One of the most intriguing comparison cases in our study is Genentech, fascinating for its squandered promise during its early years and equally interesting for its resurgence under a little-known cancer researcher promoted from within, Arthur Levinson, who instilled a 20 Mile March discipline. During its early history, Genentech pursued a strategy of breakthrough innovation—but without discipline—beginning life as the Next Big Thing incarnate, and becoming the first pure biotechnology company in history and the first to go public. It bioengineered a growth hormone for children and treatments for hairy-cell leukemia, cystic fibrosis, hemophilia, and blood clots in heart-attack patients, just to note a few of its pioneering creations. Of the heart-attack drug, the chairman of the department of medicine at Harvard Medical School said, "t-PA will do for heart attacks what penicillin did for the treatment of infections." The Next Big Thing, indeed! Yet even with all this innovation, Genentech's performance lagged behind its promise. If you'd bought Genentech stock on October 31, 1980, and held your stock until mid-1995, your investment wouldn't have even kept pace with the general stock market.[42]

Then Genentech got an incredible stroke of good fortune in promoting Arthur Levinson from chief scientist to CEO. Despite being untested as a CEO, Levinson proved to be one of the best biotechnology executives of all time, a classic Level 5 leader who detested arrogance in any form. Combining a boyish playfulness and joyful pursuit of innovation with fanatic discipline, he focused Genentech on only product categories in which it could become best in the world with a strong economic engine. Under Levinson, Genentech finally gained traction, delivering spectacular financial performance (see adjacent diagram "Genentech: Before and During Levinson") and soundly outperforming the general stock market.[43]

GENENTECH: Before and During Levinson

Profits 1980 - 2008

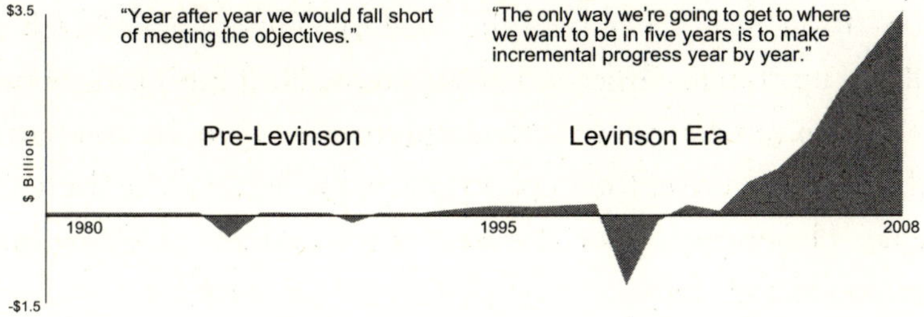

In 1998, Levinson talked openly about Genentech's historical lack of discipline, "In the past, I think we've suffered from five-year plans that represent a scenario of, 'Gee, this is what the world would look like if everything were wonderful.' And we didn't rigorously use the long range plan as a way to manage the business. Having sat through 15 of these long range planning presentations, being involved in some of them directly myself, people didn't take them seriously when you recognize that year after year we would fall short of meeting the objectives of long range plans." Then he highlighted Genentech's new approach: "The only way we're going to get to where we want to be in five years is to make incremental progress year by year . . . We've got to get 20% of the way there every year. We can't do 2% in year one, two, three and four, and 92% of it in year five. It will never happen that way."[44]

> The case of Genentech under Levinson highlights two points. First, 20 Mile Marching can help you turn underachievement into superior achievement; so long as you stay alive and in the game, it's never too late to start the march. Second, searching for—and even *finding*—the Next Big Thing does not in itself make a great company. Like a gifted but undisciplined athlete, Genentech had underperformed and disappointed, making good on its promise only once Levinson added fanatic discipline to the mix.

We live in a modern culture that reveres the Next Big Thing. It's exciting, fun to read about, fun to talk about, fun to write about, fun to learn about, and fun to join. Yet the pursuit of the Next Big Thing can be quite dangerous if it becomes an excuse for failing to 20 Mile March. If you always search for the Next Big Thing, that's largely what you'll end up doing—always searching for the Next Big Thing. The 10X cases did not generally have better opportunities than the comparisons, but they made more *of* their opportunities by 20 Mile Marching to the extreme. They never forgot: the Next Big Thing just might be the Big Thing you already have.

Of course, there remain some unanswered questions. How *do* you balance the need for fanatic discipline against the need for innovation and adaptation, especially in a chaotic world? If you just 20 Mile March, don't you run the risk of blindly marching to oblivion? How do you gain 10X success and stay alive in a world full of disruptive change, a world that demands not just discipline but also creativity and vigilance? And it's to these questions that we next turn.

20 MILE MARCH

KEY POINTS

▶ The 20 Mile March was a distinguishing factor, to an overwhelming degree, between the 10X companies and the comparison companies in our research.

▶ To 20 Mile March requires hitting specified performance markers with great consistency over a long period of time. It requires two distinct types of discomfort, delivering high performance in difficult times and holding back in good times.

▶ A good 20 Mile March has the following seven characteristics:

1. Clear performance markers.
2. Self-imposed constraints.
3. Appropriate to the specific enterprise.
4. Largely within the company's control to achieve.
5. A proper timeframe—long enough to manage, yet short enough to have teeth.
6. Imposed by the company upon itself.
7. Achieved with high consistency.

▶ A 20 Mile March needn't be financial. You can have a creative march, a learning march, a service-improvement march, or any other type of march, as long as it has the primary characteristics of a good 20 Mile March.

CHAPTER SUMMARY

▶ The 20 Mile March builds confidence. By adhering to a 20 Mile March no matter what challenges and unexpected shocks you encounter, you prove to yourself and your enterprise that performance is not determined by your conditions but largely by your own actions.

▶ Failing to 20 Mile March leaves an organization more exposed to turbulent events. Every comparison case had at least one episode of slamming into a difficult time without having the discipline of a 20 Mile March in place, which resulted in a major setback or catastrophe.

▶ The 20 Mile March helps you exert self-control in an out-of-control environment.

▶ 10X winners set their own 20 Mile March, appropriate to their own enterprise; they don't let outside pressures define it for them.

▶ A company can always adopt 20 Mile March discipline even if it hasn't had such discipline earlier in its history, as Genentech did under Levinson.

UNEXPECTED FINDINGS

▶ 20 Mile Marchers have an edge in volatile environments; the more turbulent the world, the more you need to be a 20 Mile Marcher.

▶ There's an inverse correlation between pursuit of maximum growth and 10X success. Comparison-company leaders often pressed for maximum growth in robust times, thereby exposing their enterprises to calamity in an unexpected downturn. 10X winners left growth on the table, always assuming that something bad lurked just around the corner, thereby ensuring they wouldn't be caught overextended.

► 20 Mile Marching wasn't a luxury afforded to the 10X cases by their success; they had 20 Mile Marches in place long before they were big successes, which helped them to become successful in the first place.

ONE KEY QUESTION

► What is your 20 Mile March, something that you commit to achieving for 15 to 30 years with as much consistency as Stryker, Southwest Airlines, Intel, and Progressive?

4

FIRE BULLETS, THEN CANNONBALLS

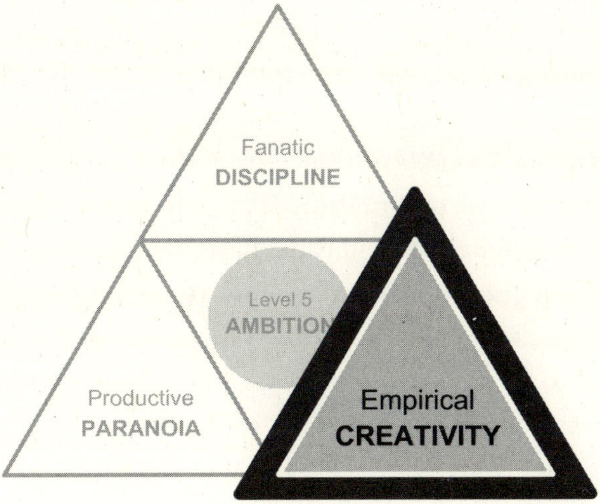

"You may not find what you were looking for, but
you find something else equally important."

—Robert Noyce[1]

Imagine you're sitting at an airline gate, waiting to board. You look up from your newspaper to see a pilot in full captain's uniform, walking to your plane . . . wearing dark glasses and tapping a white cane.

You chuckle to yourself. You've flown on this zany airline before, and you know this is just another fun trick being played on unsuspecting passengers. Pilots would sometimes leave the intercom open and say things like, "Do you remember how to start this thing?" or "I thought *you* had the keys." The airline encouraged flight attendants to engage in playful banter, invent games, and crack jokes with passengers: "We're serving steak and baked potatoes today . . . on the flight that left an hour ago." The airline singled out a passenger each time it logged another millionth customer. In one case, it acknowledged this marker by leading the tagged customer down the jet stairs and handing him the reins to a cow, a befuddled bovine standing placidly on the tarmac, as a special gift. You love this renegade airline, which had brought a radical, new model and a fun-filled, high-spirit culture to the industry.

More seriously, though, you love this airline for its low fares, consistent on-time record, and no-frills approach. Instead of having to go through the traditionally complicated ticketing procedure, you just get a simple cash-register receipt. There are no seating assignments, no first-class distinctions, and few delays. The planes land, quickly turn at the gate, and go back out again. You love the point-to-point model, with no hub-and-spoke connections. The whole experience is simple, fast, fun, reliable, safe, and cheap.

As you prepare to board the flight, hoping that you're not tagged as a millionth marker (you really don't want or need a cow), you notice one of your favorite things: the black "U" going from left to right under the front of the aircraft, creating the effect of a giant, friendly Smiley-Face staring back at you, the cockpit windows looking like eyes and the front of the aircraft taking shape as a black-tipped nose. You're taking a business trip again, on Pacific Southwest Airlines (PSA) and its giant, flying Smile Machine in the Sky.[2]

PSA became a success story for the airline industry. Not only did customers love this happy airline with its smiling aircraft, but the business model proved enormously profitable and full of growth potential. So, when a group of entrepreneurs decided to found an airline in Texas, they came up with a simple business plan: copy PSA in Texas. The *New York Times* wrote in 1971 that Southwest Airlines President Lamar Muse "says frankly—and repeatedly—that Southwest Airlines has been developed from its inception around the ideas that have proven to be successful for Pacific Southwest Airlines."[3]

"We don't mind being copycats of an operation like that," said Muse in 1971, referring to visits he and other Southwest executives made to PSA as they assembled their operating plans. PSA welcomed the Texas upstart to its San Diego operations, indeed even selling them flight and operations training. This may seem odd, but in a pre-deregulation world with Southwest constrained to Texas, PSA would remain unthreatened in its large intrastate market, California.[4]

The visiting entrepreneurs from Texas flew in PSA jump seats, and took notes on every detail of gate and backroom operations. They returned to Texas with copious notes and a set of operating manuals that they used to mimic PSA's model to the smallest detail, including the fun and zany culture. Lamar Muse later wrote that creating the operating manuals for his upstart airline was "primarily a cut-and-paste procedure," a detail corroborated by another book written on the rise and fall of PSA. Southwest Airlines copied PSA so completely that you could almost call it a photocopy![5]

A BIG SURPRISE

When we began this research effort, we anticipated that innovation might be a primary distinguishing factor for 10X success in unstable environments characterized by rapid change. But then how do we explain PSA and Southwest Airlines? Imagine our surprise to discover that the true innovator, PSA, no longer even exists as an independent brand, despite having created one of the most successful airline business models of the 20th century.[6] And further, that Southwest Airlines, one of our most favorite and beloved cases, had in fact hardly innovated anything at its founding.

We analyzed Southwest versus PSA first in this study, and we commented in our research-team discussions, "Well, perhaps airlines present a special case, wherein scale and costs count much more than innovation." Surely, we thought, when we look into technology-driven industries like medical devices, computers, semiconductors, software, and biotechnology, we'll see overwhelming evidence of the 10X companies out-innovating the comparisons.

Well, we were surprised by what we found.

Our biggest shock came when we studied our pair of companies in biotechnology, the one industry in which the correlation between innovation and success should be close to 100 percent. Take a look at the two sets of curves in the chart "Role Reversal in Biotechnology: Amgen vs. Genentech." On the left, we see Genentech's stunning performance in creative output, outpacing Amgen by more than two times in patent productivity; on the right, we see Amgen's spectacular financial performance, blowing Genentech away by a factor of more than thirty to one. Professor Jasjit Singh, who has systematically studied patent productivity, found a similar pattern in patent *citations*, showing that Genentech created not only a lot of patents but also highly *impactful* patents. Genentech stood out as one of the most innovative companies in the history of the biotechnology industry, being the first to apply recombinant DNA to a major commercial product, the first to create an FDA-approved biotechnology product, the company that *Science* magazine touted as having an unparalleled record in the industry at creating ma-

Role Reversal in Biotechnology: Amgen vs. Genentech

Innovation and Performance

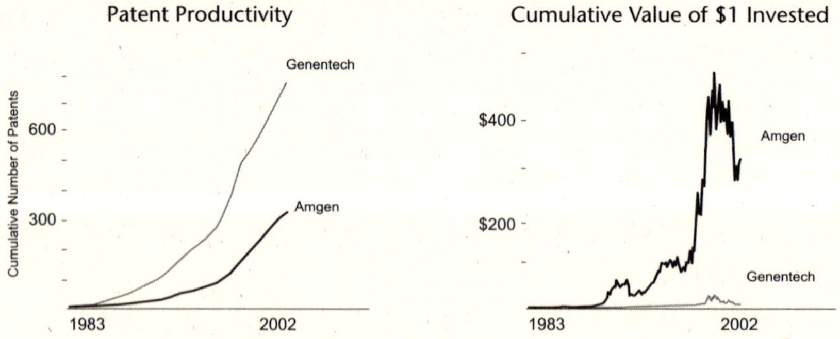

Notes:
1. Source for patent numbers in this work: United States Patent and Trademark Office, www.uspto.gov.
2. Source for all stock return calculations in this work: ©200601 CRSP®, Center for Research in Security Prices. Booth School of Business, The University of Chicago. Used with permission. All rights reserved. www.crsp.chicagobooth.edu.

jor new breakthroughs. Yet Amgen, not Genentech, became the 10X case in our study.[7]

Curious, we undertook a systematic analysis of innovation, focusing on the relevant dimensions of innovation for each industry (e.g., innovation in biotechnology focuses on new products and scientific discoveries, whereas innovation in airlines focuses on new business models and operating practices, and so forth). We identified incremental, medium, and major innovations, counting 290 innovation events (31 major, 45 medium, and 214 incremental), comparing the 10X winner to each comparison case and asking ourselves which company was more innovative during its era of analysis. (See *Research Foundations: Innovation Analysis.*) In only three of seven pairs, the 10X case proved more innovative than the comparison company.

> The evidence from our research does not support the premise that 10X companies will necessarily be more innovative than their less successful comparisons. And in some surprise cases, such as Southwest Airlines versus PSA and Amgen versus Genentech, the 10X companies were *less* innovative than the comparisons.

John Brown at Stryker lived by the mantra that it's best to be "one fad behind," never first to market, but never last. In contrast, Leon Hirsch at comparison case USSC piled breakthrough upon breakthrough, with new products that revolutionized surgical practice such as absorbable surgical staples and special devices for minimally invasive procedures, building a reputation among business analysts as the most innovative leader in its product categories. *Investor's Business Daily* remarked, "That's how [USSC] kept the competition at bay—by out-innovating them." Yet Stryker—stepping along one fad behind—trounced USSC in long-term performance.[8]

Even in pairs where the 10X case did out-innovate its comparison, such as with Intel versus AMD, the evidence still does not support the idea that maximum pioneering innovation is the most essential differentiator of 10X success. At multiple junctures in its history, Intel did not have the most innovative chip in the industry. Intel lagged behind National Semiconductor and Texas Instruments in the move to 16-bit microprocessors. Some of Intel's own executives saw the Motorola 68000 as better than Intel's 8086, and then Intel was late to market with its 32-bit microprocessors. Intel also fell behind those pioneering RISC (reduced instruction set) chips and had to play catch-up. Of course, Intel did create significant innovations—*we're not saying that Intel failed to innovate*—but historical evidence shows Intel to be less of a pioneering innovator at critical junctures than most people realize.[9]

We're not the only researchers to have such findings. We came across a fascinating piece of work by Gerard J. Tellis and Peter N. Golder in their book *Will and Vision*. Tellis and Golder systematically examined the relationship between attaining long-term market leadership and being the innovative pioneer in 66 wide-ranging markets, from chewing gum to the Internet. They found that only 9 percent of pioneers end up as the final winners in a market. Gillette didn't pioneer the safety razor; Star did. Polaroid didn't pioneer the instant camera; Dubroni did. Microsoft didn't pioneer the personal computer spreadsheet; VisiCorp did. Amazon didn't pioneer online bookselling and AOL didn't pioneer online Internet service. Tellis and Golder also found that 64 percent of pioneers failed outright. It seems that pioneering innovation is good for society but statistically lethal for the individual pioneer![10]

We envisioned sharing these puzzling findings with some of the 10X leaders, imagining they might be surprised, perhaps even outraged. We pictured Bill Gates, who viewed innovation as central to Microsoft's first three decades of success, snapping at us, "That's the stupidest thing I've ever heard!"

And indeed, if we came out and said, "Innovation is bad," we could justifiably be called stupid. But that isn't our point; we're *not* saying that innovation is unimportant. Every company in this study innovated. It's just that the 10X winners innovated less than we would have expected relative to their industries and relative to their comparison cases; they were innovative *enough* to be successful but generally *not the most* innovative.

We concluded that each environment has a level of "threshold innovation" that you need to meet to be a contender in the game; some industries, such as airlines, have a low threshold, whereas other industries, such as biotechnology, command a high threshold. Companies that fail even to meet the innovation threshold cannot win. But—and this surprised us—*once you're above the threshold,* especially in a highly turbulent environment, being more innovative doesn't seem to matter very much.

THRESHOLD INNOVATION

Industry	Primary Innovation Dimension	Innovation Threshold
Semiconductors	New devices, products, and technologies	High
Biotechnology	New drug development, scientific discoveries, breakthroughs	High
Computers/ Software	New products, enhancements, and technologies	High
Medical Devices	New medical devices, application breakthroughs	Medium
Airlines	New service features, new business models and practices	Low
Insurance	New insurance products, new service features	Low

So, we have an enticing puzzle. *Why* doesn't innovation systematically distinguish the 10X winners from the comparisons, despite the widely held view that innovation is perhaps the #1 differentiating factor of success in a fast-changing world? Because, in essence, once a company meets the threshold of innovation necessary for survival and success in a given environment, it needs a *mixture* of other elements to become a 10X company—in particular, the mixture of creativity and discipline.

CREATIVITY *AND* DISCIPLINE

In 1970, a small company named Advanced Memory Systems broke the 1,000-bit memory-chip barrier and introduced a well-designed product to the market a few months before its rival, another small company named Intel. That might not sound like much of a lead, but in the early stages of a race to become an industry standard in a rapidly changing technological revolution, falling months behind can be like falling a minute behind in a four-minute-mile race. Intel crashed the clock to introduce its 1103 memory chip in late 1970. In the rushed melee, Intel slammed into a series of problems, including a glitch (caused by an excess surface charge) that could erase data. Here sat young Intel, months behind, with a memory chip that under certain conditions couldn't remember! Intel engineers worked fifty, sixty, seventy hours a week for eight months to fix the problem. "This place was a madhouse," reflected Andy Grove in 1973. "I was literally having nightmares. I would wake up in the middle of the night, reliving some of the fights that took place during the day."[11]

And yet despite all of this, Intel caught, passed, and utterly crushed Advanced Memory Systems. "We had a better design but we blew it in the marketplace," said Advanced Memory Systems' chairman. "[Intel] just bowled us over." By 1973, Intel's 1103 had become the best-selling semiconductor component in the world, used by nearly every major computer manufacturer.[12]

The reason?

Yes, innovation played a role; the 1103 proved to be a very good chip. But more telling is a motto Intel had coined for itself by 1973: "Intel Delivers."[13]

"It was our ability to deliver the parts that swung the balance in our favor," said Robert Noyce of Intel's early success.[14] Intel obsessed over manufacturing, delivery, and scale. "We want to do one good job on engineering," continued Noyce, "and sell it over and over again."[15]

> "Intel Delivers" explains Intel's 10X success much better than "Intel Innovates." Even more accurate, "Intel innovates to a necessary threshold, then blows everyone away—utterly, completely, fanatically, obsessively—with its ability to deliver on its innovations, at expected cost, with high reliability and great consistency." This is the essence of Intel's 10X journey.

Intel's founders believed that innovation without discipline leads to disaster. "This business lives on the brink of disaster," said Gordon Moore in 1973, referring to the tendency of overeager technologists to overpromise what they can deliver and then fail to come through with enough reliable chips at affordable cost. Indeed, the original statement of Moore's Law, written by Moore in 1965, focused not just on doubling the complexity of integrated circuits per year (the innovation element) but also doing so *at minimum cost.* Adhering to Moore's Law was a discipline game, a scale game, a systems game, not just an innovation game. As Leslie Berlin wrote about the early days of Intel in her authoritative and well-written book, *The Man Behind the Microchip,* "What Intel needed going forward was not the courage to take great leaps ahead but the discipline to take orderly steps in a controlled fashion." Andy Grove said during this era, "We have to systematize things so we don't crash our technology," in an article that compared Intel's approach to making semiconductor chips to pumping out high-tech jelly beans. Grove sought to pattern Intel not after an advanced R&D lab but—of all companies—McDonalds, keeping a hamburger box on his desk with a mock logo, McIntel. A quarter of a century after the 1103 success, Intel rearticulated its core values. And what did Intel leaders choose as the #1 core value atop the list? Not innovation or creativity, but discipline.[16]

Of course, it is not discipline alone that makes greatness, but the *combination* of discipline *and* creativity. In the vernacular of *Built to*

Last, this is a true "Genius of the AND." As one longtime friend of Herb Kelleher of Southwest Airlines put it, "What people don't understand is that Herb has the crazy creativity of the Irishman and the relentless discipline of the Prussian. You just don't get that combination very often."[17]

> The great task, rarely achieved, is to blend creative intensity with relentless discipline so as to amplify the creativity rather than destroy it. When you marry operating excellence with innovation, you multiply the value of your creativity. And that's what 10Xers do.

Our data on comparative innovativeness led us to a crux dilemma. On the one hand, when you're faced with an uncertain and unstable world, an obsessive focus on innovation by itself does not make for great success and might even lead to demise; bet big on the wrong innovations or fail to execute on the right innovations, and you leave yourself exposed. On the other hand, *if you just sit still and never do anything bold or new, the world will pass you by*, and you'll die from that instead. The solution to this dilemma lies in replacing the simplistic mantra "innovate or die" with a much more useful idea: *fire bullets, then fire cannonballs.*

BULLETS, THEN CANNONBALLS

Picture yourself at sea, a hostile ship bearing down on you. You have a limited amount of gunpowder. You take all your gunpowder and use it to fire a big cannonball. The cannonball flies out over the ocean . . . and misses the target, off by 40 degrees. You turn to your stockpile and discover that you're out of gunpowder. You die.

But suppose instead that when you see the ship bearing down, you take a little bit of gunpowder and fire a bullet. It misses by 40 degrees. You make another bullet and fire. It misses by 30 degrees. You make a third bullet and fire, missing by only 10 degrees. The next bullet hits—

ping!—the hull of the oncoming ship. *Now,* you take all the remaining gunpowder and fire a big cannonball along the same line of sight, which sinks the enemy ship. You live.

On April 14, 1980, venture capitalist William K. Bowes and scientist Winston Salser brought a small group of scientists and investors to a meeting at the California Institute of Technology to discuss a newly incorporated biotechnology company. The company had no CEO, product, marketing plan, or specific direction. It had little more than a scientific advisory board and a group of people willing to invest a little under $100,000 in the emerging field of recombinant DNA. The idea was simple: get the best people they could find, fund them to throw the latest recombinant-DNA technology at a range of ideas, strike upon something that would work, create a product, and build a successful company.[18]

Six months later, Bowes convinced George Rathmann to leave his position as a vice president for R&D at Abbott Laboratories to lead this small start-up that would become Amgen. Rathmann and three employees started work in a prefab, tilt-up building shared with an evangelical choir in Thousand Oaks, California. Task 1: Get great people. Task 2: Assemble as much gunpowder (additional funding) as possible. Task 3: Find a path to success and build a great company.[19]

But how?

Amgen embraced recombinant-DNA technology and "tried it on virtually everything."[20] Amgen began firing bullets, lots of bullets:

Bullet: Leukocyte interferon, for viral diseases.
Bullet: Hepatitis-B vaccine.
Bullet: Epidermal growth factor, for wound healing and gastric ulcers.
Bullet: Immunoassays, to improve medical-diagnostic tests.
Bullet: Hybridization probes, for diagnostics in cancer, infectious disease, and genetic disorders.
Bullet: Erythropoietin (EPO), for treating anemia in chronic kidney disease.

Bullet: Chicken growth hormone, to build better chickens.
Bullet: Bovine growth hormone, to get more milk from cows.
Bullet: Growth-hormone-releasing factors.
Bullet: Porcine-parvovirus vaccine, to increase reproductive rates in pigs.
Bullet: Transmissible-gastroenteritis-virus vaccine, for intestinal infections in piglets.
Bullet: Bioengineered indigo to dye blue jeans.[21]

By 1984, erythropoietin (a glycoprotein that stimulates red-blood-cell production, used to treat anemia) began to show the most promise. As the science progressed and Amgen scientists isolated the EPO gene, Amgen allocated more gunpowder, moving into clinical trials, proving efficacy, securing a defensible patent, and so on. Then, with the science done and the market assessed (200,000 chronic-kidney-disease patients in the United States), Amgen fired a cannonball, building a testing facility, allocating capital to manufacturing, and assembling a launch team. EPO became the first super-blockbuster bioengineered product in history.[22]

> Amgen's early days illustrate a key pattern we observed in this study: fire bullets, then fire cannonballs. First, you fire bullets to figure out what'll work. Then once you have empirical confidence based on the bullets, you concentrate your resources and fire a cannonball. After the cannonball hits, you keep 20 Mile Marching to make the most of your big success.

The history of the 10X companies is like a battlefield pockmarked with craters, and littered with bullets that never hit anything and lodged themselves in the ground. (See the following table, "What Makes a Bullet?") Retrospective accounts tend to focus on only the big cannonballs, giving the false impression that 10X achievements come to those with the guts to go always for the big bet, the huge cannonball. But the historical research evidence presents a different story, a story of dozens of small bullets that thumped into the dirt, punctuated by a handful of cannonballs that smashed into their targets.

WHAT MAKES A BULLET?

A bullet is an empirical test aimed at *learning what works* and that meets three criteria:

1. A bullet is *low cost.* Note: the size of a bullet grows as the enterprise grows; a cannonball for a $1 million enterprise might be a bullet for a $1 billion enterprise.

2. A bullet is *low risk.* Note: low risk doesn't mean high probability of success; low risk means that there are minimal consequences if the bullet goes awry or hits nothing.

3. A bullet is *low distraction.* Note: this means low distraction for the overall enterprise; it might be very high distraction for one or a few individuals.

10X companies used a combination of creative bullets (such as new products, technologies, services, and processes) and acquisitions. For an acquisition to qualify as a bullet, it needs to meet the three tests: low cost, low risk, and low distraction. Biomet used acquisitions to explore new markets, technologies, and niches but did so with a self-imposed constraint. Acquisitions would be made with little or no debt, and only when the balance sheet would remain strong after the purchase, thereby ensuring that acquisitions would remain low risk, low cost, and relatively low distraction.[23]

In contrast, Kirschner, Biomet's comparison case, made *cannonball acquisitions,* taking on significant debt and risk. (See diagram "Biomet vs. Kirschner.") Kirschner's acquisitions *had* to hit the target, else the company would be in serious trouble. In 1988, Kirschner made a cannonball acquisition of Chick Medical at a price exceeding 70 percent of

Kirschner's total stockholders' equity.[24] It turned out to be a disastrous move, made worse when Chick Medical's sales force defected to a competing firm. As Kirschner financed this and other acquisitions, its ratio of total liabilities to stockholders' equity skyrocketed from 43 percent to 609 percent, leaving the company terribly exposed. Bleeding cash, crushed by debt, its huge cannonball acquisitions having achieved little, Kirschner was forced to sell out in 1994 to Biomet.[25]

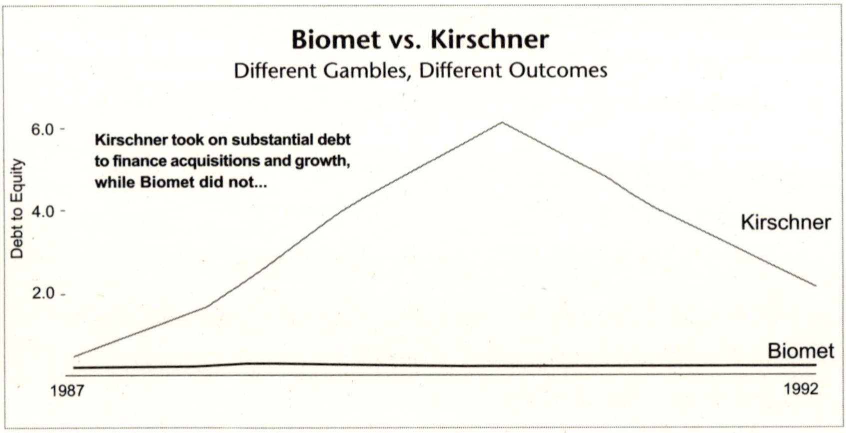

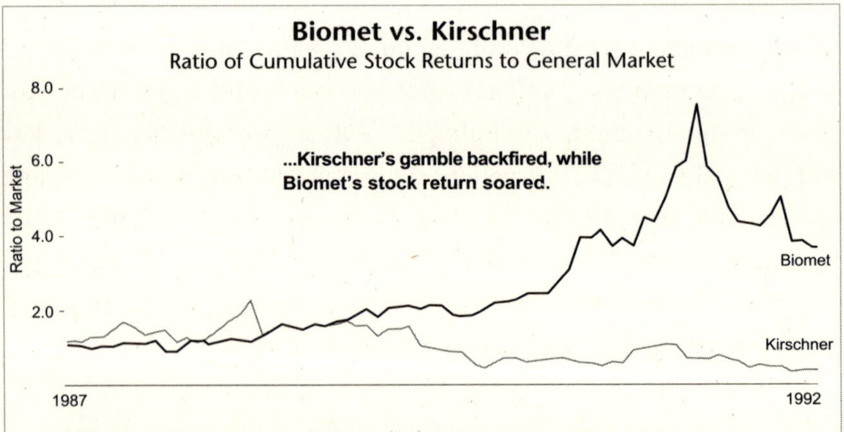

Notes:
1. Debt = total current liabilities + total long-term liabilities.
2. Each company's ratio to market was calculated from December 31, 1986 to December 31, 1992
3. Source for all stock return calculations in this work: ©200601 CRSP®, Center for Research in Security Prices. Booth School of Business, The University of Chicago. Used with permission. All rights reserved. www.crsp.chicagobooth.edu.

THE DANGEROUS LURE OF UNCALIBRATED CANNONBALLS

Embracing the "fire bullets, then cannonballs" principle requires a combination of activities:

- Fire bullets.
- Assess: Did your bullets hit anything?
- Consider: Do any of your successful bullets merit conversion to a big cannonball?
- Convert: Concentrate resources and fire a cannonball *once* calibrated.
- Don't fire uncalibrated cannonballs.
- Terminate bullets that show no evidence of eventual success.

Both the 10Xers and the comparison cases fired cannonballs. The comparison companies, however, tended to fire cannonballs *before* they'd obtained a confirming calibration—empirical validation gained through actual experience—that the cannonball would likely reach its intended target. For shorthand, we call a cannonball fired before you gain empirical validation an *uncalibrated cannonball*. The 10Xers were much more likely to fire *calibrated cannonballs*, while the comparison cases had uncalibrated cannonballs flying all over the place (the 10X cases had a 69 percent calibration rate on cannonballs versus 22 percent for the comparisons). Whether fired by the 10X case or the comparison case, calibrated cannonballs had a success rate nearly four times higher than uncalibrated cannonballs, 88 percent to 23 percent. (See *Research Foundations: Bullets-Then-Cannonballs Analysis*.)

In 1968, PSA launched a bold new cannonball called "Fly-Drive-Sleep." On the surface, the idea made sense. You're an airline. People who fly need to rent cars and they need hotel rooms. So, move into the hotel and rent-a-car businesses. PSA began buying and taking out 25-year leases on California hotels, including the permanently docked ocean liner the *Queen Mary*. It also bought a rental-car company, rapidly expanding to 20 locations and more than two thousand cars. PSA could have fired a series of bullets, buying one hotel, partnering with a rental-car company, testing it out in a specific location, learning where it could (and could not) make the concept work. Instead, it went big, and unfortunately, the Fly-Drive-Sleep cannonball flew off into the ether, generating losses every single year. "We're damn poor hotel operators," reflected PSA Chairman J. Floyd Andrews.[26]

Then in the early 1970s, PSA fired another uncalibrated cannonball when it contracted to buy five L1011 super-wide-body jumbo jets at a price equal to 1.2 times its total stockholders' equity. Keep in mind, PSA was a short-haul commuter, doing rapid gate turns to shuttle people up and down the California corridor (not a great fit with super-wide jumbo jets that can take a long time to board). Furthermore, PSA made special modifications (such as wider exit doors and no food-preparation galley), which would make the jets hard to sell to other airlines if PSA needed cash. The L1011 plan required substantial upfront investment in new towing tractors, maintenance equipment, boarding equipment, and training. Forty-two-thousand-pound thrust engines would burn through gigantic quantities of jet fuel, inflicting huge per-flight losses if PSA failed to fill the 302-seat aircraft.[27]

Unfortunately, an Arab oil embargo doubled jet-fuel prices just as PSA began to put the huge L1011s into service and struggled to exit from its Fly-Drive-Sleep fiasco. The economy fell into recession. Inflation drove up costs, yet the California Public Utilities Commission (which regulated airline prices) granted only a 6.5 percent fare increase in response to PSA's plea for 16 percent. Then the machinists' union went on strike. The L1011s went unfilled, and eventually they were mothballed in the desert, never to fly again with the PSA fleet. Said PSA's senior vice president for finance in 1975, "We have come very, very close to insolvency."[28]

PSA never regained its prior greatness and continued to fire uncalibrated cannonballs in a desperate attempt to regain momentum. It tried to launch a joint venture with Braniff Airlines, hoping for a shortcut to becoming a national carrier (the potential venture ended when Braniff went bankrupt); abandoned its simple no-frills model; switched to McDonnell Douglas aircraft for its smaller jets (moving away from its proven success with Boeing); and moved into the oil-and-gas-exploration business. And it did all this while getting clobbered by a never-ending series of disruptive events. Deregulation exposed PSA to a swarm of ferocious competitors. A lawsuit with Lockheed over the L1011s created financial uncertainty. A pilots' strike shut down the airline for 52 days. A shift to McDonnell Douglas DC-9-80s came with unexpected delivery delays, leaving PSA short of aircraft just as the strike ended, dashing the airline's reputation for reliable, on-time performance. And tragically, a Cessna student-trainer airplane hit a PSA 727 descending into San Diego, sending both aircraft hurtling to the ground. "Tower, we're going down," said the jet pilot. "This is PSA."[29]

Finally, on December 8, 1986, PSA capitulated to a buyout from US Air. PSA jets with the signature smile rolled one by one into yawning hangers, where workmen attacked them with chemicals and blasters. The aircraft emerged, faceless, repainted as interchangeable machines in a giant fleet.[30]

> PSA's demise illustrates the danger of firing uncalibrated cannonballs in an uncertain world full of turbulent disruption. If an enterprise gets slammed by a series of shocks just as its uncalibrated cannonballs go crashing off into space, it's more likely to have a catastrophic outcome.

Of course, we're focusing here on uncalibrated cannonballs that don't find a target. But what if you fire uncalibrated cannonballs that do hit a target? If there's a big enough potential payoff, perhaps the big uncalibrated bet is worth the risk. But here's the irony: firing an uncalibrated cannonball that succeeds, generating a huge windfall, can be even *more* dangerous than a failed cannonball. Keep in mind the danger of achieving good outcomes from bad process. Good process doesn't guarantee good outcomes, and bad process doesn't guarantee bad outcomes, but good outcomes with bad process—firing uncalibrated cannonballs that just happen to succeed—reinforces bad process and can lead to firing more uncalibrated cannonballs.

Would you advise a friend or relative to go to Las Vegas and bet half of his entire net worth on a single spin of the roulette table? Suppose your friend believes that people win big only if they make big risky bets on games like roulette, and he heads off to Vegas, places a huge roulette bet, and wins. He comes home and says, "See, it's a good idea to bet on roulette, just look at my success. I'm going back next week to bet my *entire* net worth!"

10XERS LEARN FROM THEIR FOLLIES

The 10X cases didn't have perfect records in calibrating their cannonballs. Southwest bought Muse Air in the early 1980s, a big move outside its proven model; it failed. Intel made an uncalibrated bet in the 1990s to push the personal computer industry to a new memory technology from RAMBUS; it failed. But in the rare instances in which the 10X cases fired uncalibrated cannonballs, they quickly learned from their mistakes and returned to a bullets-then-cannonballs approach.[31]

For most of its history, Progressive Insurance lived by an explicit guideline to prevent uncalibrated cannonballs: limit any new business to 5 percent of total corporate revenues until fine-tuned for sustained profitability. Progressive broke this rule in the mid-1980s when it moved into selling insurance to trucking companies and transit-bus systems, jumping from zero to $61 million in net premiums written (almost 8 percent of total Progressive premiums) in less than two years. It multiplied the trucking-insurance staff nearly ten times in a single year—despite an underwriting loss of 23 percent—and then nearly tripled premiums again the next year. "We thought the market was just bad drivers with bigger cars," said a Progressive executive. But the business turned out to be very different; trucking companies had much greater power to negotiate prices than individual drivers, and they had armies of sophisticated lawyers to battle claims' disputes. A "financial disaster," said Lewis of the $84 million loss that followed. "I'm ashamed for how we got into that position," he admitted. Then he pointed in the mirror to apportion blame: "I truly am responsible for that."[32]

> Even 10Xers make mistakes, even sometimes the big mistake of firing an uncalibrated cannonball. But they view mistakes as expensive tuition: better get something out of it, learn everything you can, apply the learning, and don't repeat. Whereas comparison cases often try to recover from the calamity of firing an uncalibrated cannonball by firing yet another uncalibrated cannonball, 10Xers recover by returning to the discipline of firing cannonballs only when they have empirical validation.

Progressive vowed never to make the uncalibrated-cannonball mistake again and subsequently applied the lesson in its move into standard insurance. Progressive had built its success primarily upon non-standard insurance, selling to high-risk drivers shunned by traditional insurers. Should Progressive move into standard insurance, selling to the broad spectrum of drivers? Progressive executives didn't know, but they knew how to find out: fire bullets.[33]

In 1991, Progressive crafted experiments in a handful of states it knew well, such as Texas and Florida. Two years later, it continued firing bullets, testing standard insurance in more states. Bullet, bullet, bullet . . . each one showed results, each one validated the concept. Then in 1994, with empirical validation—we've proven we can do this!—Progressive concentrated a whole bunch of gunpowder, firing a cannonball, committing fully to standard insurance. By the end of 1996, Progressive offered standard insurance in all 43 states where it operated. Within five years, standard insurance accounted for nearly half of Progressive's overall business, eventually catapulting it to the #4 spot overall in the American auto-insurance industry by 2002.[34]

In an interesting contrast to both the uncalibrated trucking cannonball and the calibrated standard-auto-insurance cannonball, Progressive decided *not* to fire a cannonball on homeowners insurance. At first glance, the idea of selling homeowners insurance made sense. After all, why not enable customers to bundle together car and home insurance? We can envision reams of analysis demonstrating the synergies and strategic rationale for such a move, perhaps even making the case for a giant acquisition. But Progressive had learned: you can only know if something will *actually work* if you gain empirical validation, no matter how many slide decks support the idea. So, Progressive turned again to bullets, just like the move into standard auto insurance, testing in a handful of states. However, unlike the bullets fired into standard auto insurance, the homeowners-insurance bullets hit nothing, and Progressive pulled the plug.[35]

> Progressive's three strategic decisions—trucking insurance (uncalibrated cannonball), standard auto insurance (calibrated cannonball), and homeowners insurance (bullets followed by the decision not to fire a cannonball)—all underscore one very big lesson. In the face of instability, uncertainty, and rapid change, relying upon pure analysis will likely not work, and just might get you killed. Analytic skills still matter, but empirical validation matters much more.

And that's the underlying principle: *empirical validation*. Be creative, but validate your creative ideas with empirical experience. You don't even need to be the one to fire all the bullets; you can learn from the empirical experience of others. Southwest Airlines became one of the most successful start-up companies of all time by betting on an empirically validated model that it copied from PSA. Roald Amundsen built his strategy on proven techniques, such as the use of dogs and sleds, that'd been honed for centuries by Eskimos. (Robert Falcon Scott, in contrast, bet big on his newfangled motor sledges, which had never been fully tested in the most extreme polar conditions.) More important than being first or the most creative is figuring out *what works in practice*, doing it better than anyone else, and then making the very most of it with a 20 Mile March.[36]

EMPIRICAL VALIDATION, NOT PREDICTIVE GENIUS

When we began this research study, we wondered whether the 10X winners would prove to be superior at predicting the future, putting themselves ahead of the curve and winning big because of their predictive genius. But we didn't find this to be true. Even the great software genius Bill Gates had no special predictive ability. He didn't plan from the outset for Microsoft to be first to the market with an operating system for the IBM PC; he was off focusing on computer languages when IBM unexpectedly asked if Microsoft could provide an operating system. Nor did he lead Microsoft to be first in the Internet-browser market.[37]

In 1987, Bill Gates faced a conundrum, whether to bet on DOS/Windows or OS/2. On the one hand, the IBM PC had become a standard based on MS-DOS, and Microsoft had written Windows to run on DOS, giving Windows an early standards advantage. On the other hand, IBM had made a huge commitment to building a new operating system and had engaged Microsoft in developing what would become known as OS/2. In April 1987, IBM stormed the industry with its new line of computers running the technically superior OS/2, and Gates himself predicted that within two years OS/2 would dominate.[38]

Yet at the exact same time, without fanfare, Gates also fired bullets on continued Windows development. After all, what if OS/2 failed? What if the DOS standard proved too big to overcome, even for IBM? What if software companies didn't convert their programs to run on OS/2, leaving the new computers without a wide range of software options? *What if? What if? What if?* Exercising his productive paranoia, Gates worried about leaving Microsoft exposed to all these uncertainties, and so, despite vigorous challenges from some in his own inner circle, he hedged by keeping a handful of people on Windows . . . just in case. Gates was smart enough to know that he wasn't smart enough to predict with certainty what would actually happen to OS/2.[39]

By late 1988, OS/2 had garnered only 11 percent of the market. Bad news for IBM but not necessarily for Microsoft, as *Business Week* put it, "In a way, Microsoft can't lose. Should OS/2 falter, MS-DOS will pick up the slack." Gates continued to forecast, publicly at least, that OS/2 would win. But the empirical evidence began to turn in Windows' favor. "Who would have predicted . . . that 1989 would be the Year of Microsoft Windows, rather than of OS/2?" wrote *PC Week*. "Yet that seems to be the case." Windows 3 hit the market and sold a million copies in just four months, compared to just three hundred thousand copies of OS/2 in three years.[40]

So, Gates bet fully on Windows. By 1992, Windows was selling more than a million copies per month, and Gates then committed to building Windows 95. The cannonball smashed into its target, with Windows 95 reaching a million customers within four days, giving Microsoft a dominant position. Microsoft just kept on going, 20 Mile Marching, making the most of a Very Big Thing.[41]

> 10Xers do not have any particular genius for visionary prediction. If Bill Gates, one of the great business geniuses of the 20th century, couldn't accurately predict what was going to happen in his environment, there's little reason to expect that *anyone* can succeed with a "predict the future and then position yourself for what's coming" strategy.

We were relieved to discover that you don't need any special predictive ability to thrive in uncertainty. If you don't know what's going to happen next—and no one does—this chapter outlines a method for making progress rather than getting paralyzed, frozen by life's uncertainties. As we progressed in our work and learned how the 10Xers dealt with uncertainty and change, we began to change our own approach, even our own terminology, moving away from trying to predict the future or to analyze our way to the "right" answer. Instead, we began to ask questions like:

"How can we bullet our way to understanding?"
"How can we fire a bullet on this?"
"What bullets have others fired?"
"What does this bullet teach us?"
"Do we need to fire another bullet?"
"Do we have enough empirical validation to fire a cannonball?"

If you knew ahead of time which bullets would merit cannonballs, you'd fire only those. But of course, you don't know, so you need to fire bullets, knowing full well that a number of them will never hit anything. Eventually, however, there comes a time for commitment, when you have enough validation to fire the cannonball; if you fire only bullets but never commit to a big bet or an audacious objective, you'll never do anything great.

APPLE'S REBIRTH: BULLETS, CANNONBALLS, AND DISCIPLINED CREATIVITY

When Steve Jobs decided to move Apple into retail stores in the early 2000s, he understood that he didn't know how to do it. Lacking empirical experience, he asked, "Who is the best retail executive?" The answer: Mickey Drexler, then CEO of The Gap. So, Jobs lured him onto Apple's board and began learning everything he could. Drexler told Jobs not to just launch with a big roll-out of twenty or forty stores. Instead, go off to a warehouse, prototype a store, redesign it until you have it right (bullet, bullet, bullet), and roll it out to the world (cannonball) only once you've got it working and tested. That's exactly what Jobs did. And indeed, the first iteration just didn't work: "We were like, 'Oh God, we're screwed,' " said Jobs. So, Jobs and his retail leader, Ron Johnson, redesigned, tested, and redesigned until they got it right. They launched their first two stores in Virginia and Los Angeles, and once those proved successful, they rolled them out with great consistency. Bullet, calibrate, bullet, recalibrate, cannonball.[42]

Steve Jobs had returned to Apple in 1997, having wandered in the high-tech wilderness for 12 years after losing a boardroom showdown with John Sculley, the CEO whom Jobs had brought in to help him run the company in the early 1980s. Imagine the outrage of being forced out of your own company, then watching it languish and stumble under a series of CEOs who just didn't understand what had made the company great in the first place, its cumulative stock returns falling more than 60 percent behind the general market. By the time Jobs returned, few gave Apple much hope of a return to greatness. When asked what he'd do with Apple, Michael Dell, founder of Dell Computers, told an audience at the Gartner Symposium ITxpo97, "What would I do? I'd shut it down and give the money back to the shareholders."[43]

Over the subsequent five years, from the end of 1997 through 2002, Apple outperformed the general stock market by 127 percent and then just kept going, eventually becoming the most valuable technology company in the world in 2010.

What did Jobs first do to get Apple back on track? Not the iPod, not iTunes, not the iPhone, not the iPad. First, he increased discipline. That's right, discipline, for without discipline there'd be no chance to do creative work. He brought in Tim Cook, a world-class supply-chain expert, and together Jobs and Cook formed a perfect yin-yang team of creativity and discipline. They cut perks, stopped funding the corporate sabbatical program, improved operating efficiency, lowered overall cost structure, and got people focused on the intense "work all day and all of the night" ethos that'd characterized Apple in its early years. Overhead costs fell. The cash-to-current-liabilities ratio doubled, and then tripled. Long-term debt shrunk by two-thirds and the ratio of total liabilities to shareholders' equity dropped by more than half from 1998 to 1999. [44] Now, you might be thinking, "Well, all that financial improvement naturally follows breakthrough innovation." But in fact, Apple did all this *before* the iPod, iTunes, or the iPhone. Anything that didn't help the company get back to creating great products that people loved would be tossed, cut, slashed, and ruthlessly eliminated.

What products did Apple work on first? It went backward in time to resurrect the biggest thing that Steve Jobs had helped create more than a decade earlier, the Big Thing of tremendous value still in the mix: the Macintosh personal computer. Apple launched PowerMacs, PowerBooks, and the iMac. Jobs didn't first go after the Next Big Thing, but instead he made the most of the Big Thing he already had.

Then, fully *four years after* Jobs returned to Apple, came a small, empirically validated shot. While Apple focused on the Mac, something happened entirely outside Apple's walls: music file-sharing on Napster and the introduction of MP3 digital-music players. Jobs told Brent Schlender of *Fortune* that he "felt like a dope" for being caught totally off guard by the rise of Napster, digital-music file-sharing, and MP3 players. "I thought we had missed it," Jobs continued. "We had to work hard to catch up."[45]

Consider all the empirical facts in place *before* Apple began to develop the iPod. Young people shared music; MP3 players allowed them to take their music with them anywhere; MP3 players had limited capacity; Apple had an uncanny ability to make technology accessible to "the rest of us"; a really cool MP3 player that worked with the Mac would further extend the Mac; Apple employees wanted a cool MP3 player and music library for themselves; and much of the technology needed to build a better MP3 device already existed (small hard drives from Toshiba, miniature batteries from Sony, FireWire interface from Texas Instruments, and MP3-hardware blueprint from PortalPlayer).[46]

So, Apple pulled together a nifty MP3 player for the Mac, along with supporting software, but it didn't create a Giant Leap Forward. Apple itself didn't seem to view the iPod as a significant new product category but really more of an extension. Apple's 2001 Form 10-K described the iPod as simply "an important and natural extension of Apple's digital hub strategy" based on the Macintosh personal computer—no revolution yet; just an evolutionary step in an existing strategy. By 2002, the iPod remained a small part of Apple's overall portfolio, accounting for less than 3 percent of net sales, meriting neither a separate line item in Apple's financial statements nor a mention in the opening paragraph of the company's business description. The iPod was a very cool bullet, but a bullet nonetheless.[47]

Still, Apple had increasing empirical validation. People loved the iPod; customers loved iTunes for the Mac; iPod sales more than doubled in a year; the music industry faced severe challenges from growth in illegally downloaded music; and Apple employees wanted an easy way to download music without stealing.

So, Apple took the next step, launching an online music store and working out a deal with the music industry to offer individual songs at 99 cents. This, too, succeeded, and Apple had more empirical validation. Millions of people would rather buy music than steal it, if easy to access and fairly priced; people were clamoring for iTunes for their Windows-based personal computers; and Windows had an installed base of more than one billion personal computers.[48]

Finally, with all this empirical validation, Apple fired the big cannonball, iTunes and iPod for non-Mac computers, instantly multiplying the potential market by nearly twenty times.[49] "The iPod is not a new category," said Jobs. "It's not a speculative market. . . . So it's not like saying we're going to go build an information appliance or some technical curio and hope the market exists." And it didn't stop there. Apple kept adding piece upon piece, making the most of the new Big Thing: iPod Mini, iPod Click Wheel, iPod Photo, iPod 30GB, iPod 60GB, iPod 80GB, iPod Shuffle, iPod Nano, along with movies, videos, books, and television shows at the iTunes store. Within three years, iPod unit sales would exceed Macintosh unit sales.[50]

The iPod story illustrates a crucial point: a big, successful venture can look *in retrospect* like a single-step creative breakthrough when, in fact, it came about as a multistep iterative process based more upon empirical validation than visionary genius. The marriage of fanatic discipline and empirical creativity better explains Apple's revival than breakthrough innovation per se.

The same point holds for Steve Jobs himself. When banished to the high-tech wilderness in 1985 after being ousted from his own company, Jobs never stopped developing, growing, learning, pushing himself. He could have taken his fortune, and retired to a life of ease and comfortable irrelevance. Instead, he launched a new company called NeXT, worked on a new operating system, and became engaged with animated films at Pixar. In the 12 years away from Apple, Jobs had turned himself from a creative entrepreneur into a disciplined, creative company builder. Jobs always knew how to build insanely great products, but he had to learn how to build an insanely great company.

Fanatic discipline *and* empirical creativity—two sides of a coin, both required for 10X success and enduring greatness. Still, they are not enough, for if you get knocked out of the game, all your creativity and discipline amount to nothing. Apple nearly disappeared as an independent company in the mid-1990s, having fallen so far and become so dispirited that its leaders seriously entertained a sellout to another company. Apple got a stay of execution when its board couldn't come to terms with the potential acquirers, and Jobs returned soon thereafter.[51] If Apple had capitulated and been acquired, there'd very likely have been no iMac, iPhone, iPod, or iPad. Greatness requires the Churchillian resolve to never give in, but it also requires having the reserves to endure staggering defeats, bad luck, calamity, chaos, and disruption. In a stable and predictable world, leading with fanatic discipline and empirical creativity might be enough; but uncertainty and instability also require leading with productive paranoia, the subject of our next chapter.

CHAPTER SUMMARY

FIRE BULLETS, THEN CANNONBALLS

KEY POINTS

▶ A "fire bullets, then cannonballs" approach better explains the success of 10X companies than big-leap innovations and predictive genius.

▶ A bullet is a low-cost, low-risk, and low-distraction test or experiment. 10Xers use bullets to empirically validate what will actually work. Based on that empirical validation, they then concentrate their resources to fire a cannonball, enabling large returns from concentrated bets.

▶ Our 10X cases fired a significant number of bullets that never hit anything. They didn't know ahead of time which bullets would hit or be successful.

▶ There are two types of cannonballs, calibrated and uncalibrated. A calibrated cannonball has confirmation based on actual experience—empirical validation—that a big bet will likely prove successful. Launching an uncalibrated cannonball means placing a big bet without empirical validation.

▶ Uncalibrated cannonballs can lead to calamity. The companies in our research paid a huge price when big, disruptive events coincided with their firing uncalibrated cannonballs, leaving them exposed. Comparison cases had a much greater tendency to fire uncalibrated cannonballs than the 10X cases.

▶ 10Xers periodically made the mistake of firing an uncalibrated cannonball, but they tended to self-correct quickly. The comparison cases, in contrast, were more likely to try to fix their mistakes by firing yet another uncalibrated cannonball, compounding their problems.

- Failure to fire cannonballs, *once calibrated*, leads to mediocre results. The idea is not to choose between bullets or cannonballs but to fire bullets first, *then* fire cannonballs.

- Acquisitions can be bullets, if they remain low risk, low cost, and relatively low distraction.

- The difficult task is to marry relentless discipline with creativity, neither letting discipline inhibit creativity nor letting creativity erode discipline.

UNEXPECTED FINDINGS

- The 10X winners were not always more innovative than the comparison cases. In some matched pairs, the 10X cases proved to be less innovative than their comparison cases.

- We concluded that each environment has a threshold level of innovation, defined as a minimum level of innovation required even to be a contender in the game. For some industries, the innovation threshold is low, whereas for other industries, the threshold is very high. However, once above the innovation threshold, being *more* innovative doesn't seem to matter very much.

- 10Xers appear to have no better ability to predict impending changes and events than the comparisons. They aren't visionary geniuses; they're empiricists.

- The combination of creativity *and* discipline, translated into the ability to scale innovation with great consistency, better explains some of the greatest success stories—from Intel to Southwest Airlines, from Amgen's early years to Apple's resurgence under Steve Jobs—than the mythology of big-hit, single-step breakthroughs.

ONE KEY QUESTION

▶ Which of the following behaviors do you most need to increase?

- Firing enough bullets
- Resisting the temptation to fire uncalibrated cannonballs
- Committing, by converting bullets into cannonballs once you have empirical validation

5

LEADING ABOVE THE DEATH LINE

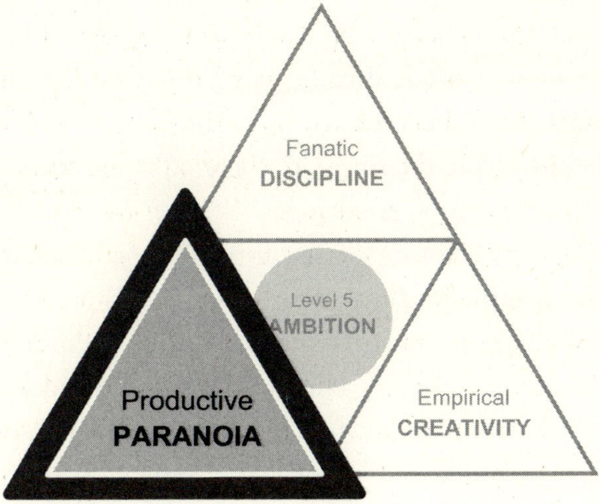

"As soon as there is life there is danger."

—Ralph Waldo Emerson[1]

On the morning of May 8, 1996, David Breashears looked down from Camp III at 24,500 feet, high on the icy slopes of Mount Everest, preparing for the big move to the South Col and a bid to carry what he called "The Pig" to the summit. The Pig was a 42-pound IMAX camera, being used to create the first-ever IMAX movie from the highest point on Earth.[2]

What Breashears saw three thousand feet below alarmed him. More than fifty people trekked out from Camp II, swarming across the glacier, climbing toward Breashears and his team. Some of the climbers were clients being led to the top of the world by experienced guides Rob Hall and Scott Fischer. Furthermore, Breashears and his team were already getting a late start, sleep deprived and on edge from hurricane-force winds that had battered their tents the night before.[3]

Breashears paused to consider: *What if* his team had to delay for a day, due to continued wind or storm, giving the swarm of climbers a chance to catch up? *What if* a bunch of people crowded the small tip-top of the mountain just as Breashears tried to film his summit shot? *What if* dozens of climbers stacked up at the bottleneck known as the Hillary Step, just before the summit, where only one climber at a time could pass up or down on fixed ropes? *What if* the combined weight of so many people weighting the fixed ropes caused anchors to rip out of the ice? *What if* the previous night's severe wind presaged a change in weather? *What if* an unexpected storm swept up the mountain like some giant bear's maw, swiping climbers off the face and sending them hurtling to their doom? *What if* he ran into a traffic jam of less experienced climbers—weakened, exhausted, disoriented—at the very moment when he needed to go down fast?[4]

Breashears had assembled the best film climbing team in the world, and he conferred with his trusted partners, Ed Viesturs and Robert Schauer. They all agreed, conditions just didn't feel right, and they came to a clear decision: Secure the gear at Camp III. Go down. Climb back up a few days later, after the mountain had cleared.[5]

On the way down, Breashears crossed paths with guide Rob Hall, tall and confident in a scarlet outfit, commanding his little army of guides and clients, moving up the mountain slowly but with almost military precision. Breashears felt a touch of chagrin, as the day had turned bright and calm, almost pleasant, and Hall looked surprised to see Breashears heading down in such great conditions. Hall looked the Master of Everest as he marched upward, while Breashears quickly passed on by and headed down toward the lower camps. Soon Breashears passed another guide, Scott Fischer, a charismatic force of energy with wild hair; a gigantic, kid-like grin; and passionate love of the mountains. Fischer, like Hall, had questions about Breashears's decision to go down, and Breashears told Fischer about the wind and questionable weather, and that the mountain felt crowded. Fischer flashed a broad, reassuring smile and continued upward, exuding his trademark optimism and joy at being on the mountain in such glorious weather.[6]

The next time Breashears would see Hall and Fischer, 15 days later, en route to his successful IMAX film shot on the summit, both Hall and Fischer would be dead, frozen high on the mountain, victims of the greatest disaster in Everest history, in which eight people had died in 24 hours.[7]

PRODUCTIVE PARANOIA

Many people know this 1996 Everest story through Jon Krakauer's book *Into Thin Air*; if you haven't yet read Krakauer's book, be sure to do so. But also be sure to read David Breashears's superb book *High Exposure*. Here, like Amundsen and Scott, we have a comparison contrast: two sets of team leaders on the same mountain on the same day, both with burdens of responsibility and business pressures (one to lead clients to the top for a large fee and the other to complete a multimillion-dollar movie project), both with tremendous experience—yet only one leads his team to 10X success, achieving the incredible goal of shooting an IMAX film on the top of Everest, and bringing himself and every member of his team safely home.[8]

It would be easy to focus on the crucial decisions made *on* the mountain. We have Breashears's prudent decision to go down on May 8, which likely saved his expedition, and perhaps even his team members' lives. Then there was Rob Hall's decision to ignore his turnaround time, not by minutes but by hours, as he waited for client Doug Hansen to reach the summit. (The "turnaround time" is a time preset by a climbing team by which they commit to begin their descent, regardless of whether they've reached the summit, thus preserving a greater margin of safety for completing their descent in daylight.) But focusing on these two moments of decision obscures our view and limits our understanding. From a 10X perspective, the most important decisions were made before the teams even got to the mountain, months before, when Breashears sat in Boston planning and preparing.[9]

David Breashears and his team brought enough oxygen canisters for more than one summit bid and enough supplies to stay at Everest for an extra three weeks. Breashears turned around and went down on May 8 because he *could* go down, wait for a better day, and still have reserves for another bid. Rob Hall's team, in contrast, brought enough oxygen for only *one* summit bid.[10] Once the guided teams set out for the summit, they'd be in a one-shot, all-or-nothing box; unlike Breashears, they didn't have the option to go down and come back on another day. When the crucial moment came high on the mountain as they ran up against the appointed turnaround time, they broke their turnaround protocol, leaving themselves terribly exposed to a fast-advancing storm and looming darkness. When the storm enveloped them, Breashears heroically gave more than half his team's canisters stored high on the mountain to help in the rescue attempts, willing to risk the multimillion-dollar film project to help save the lives of fellow climbers; even so, he was able to pull together enough resources to regroup after the tragedy and summit with the IMAX camera almost two weeks later.[11]

David Breashears's approach to Everest exemplifies the ideas in this chapter, which addresses how 10Xers lead their companies with productive paranoia. The 10X winners in our research always assumed that conditions can—and often do—unexpectedly change, violently and fast. They were hypersensitive to changing conditions, continually asking, "What if?" By preparing ahead of time, building reserves, maintaining "irrationally" large margins of safety, bounding their risk, and honing their disciplines in good times and bad, they handled disruptions from a position of strength and flexibility. They understood, deeply: *the only mistakes you can learn from are the ones you survive.*

The diagram "10X Journey and the Death Line" illustrates the idea. The rising curve represents the "10X Journey." The erratic spikes cutting across the curve represent "good events" and "bad events" that

10X Journey and the Death Line

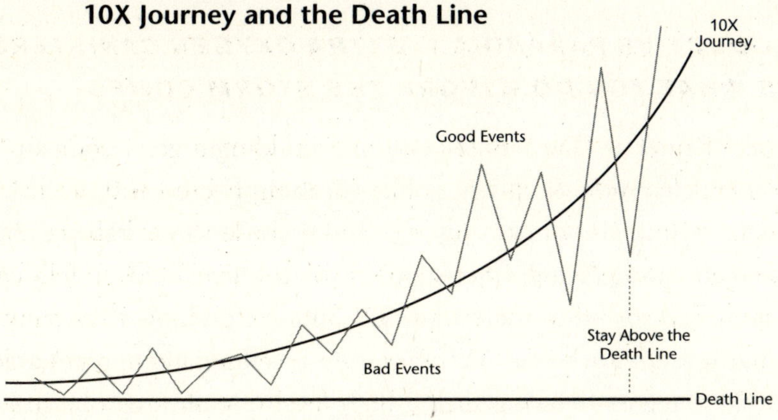

you encounter along the journey. Notice the horizontal straight line labeled "Death Line" shooting directly across the chart. In this context, "Hitting the Death Line" means that the enterprise dies outright or becomes so damaged that it can no longer continue with the quest to become an enduring great company. The idea is simple: If you ever hit the Death Line, you end the journey—game over!

In this chapter, we explore three core sets of practices, rooted in the research, for leading and building a great enterprise with productive paranoia:

- *Productive Paranoia 1:* Build cash reserves and buffers—*oxygen canisters*—to prepare for unexpected events and bad luck *before* they happen.
- *Productive Paranoia 2:* Bound *risk—Death Line risk, asymmetric risk*, and *uncontrollable risk*—and manage time-based risk.
- *Productive Paranoia 3: Zoom out*, then *zoom in*, remaining hypervigilant to sense changing conditions and respond effectively.

PRODUCTIVE PARANOIA 1: EXTRA OXYGEN CANISTERS— IT'S WHAT YOU DO BEFORE *THE STORM COMES*

Think of Intel as David Breashears and building a great company in the microelectronics industry as like climbing Everest with an IMAX camera. Think also of cash reserves and a conservative balance sheet as oxygen canisters and other supplies. By the late 1990s, Intel's cash position had soared to more than $10 billion, reaching 40 percent of annual revenues (whereas AMD's cash-to-revenue ratio hovered at less than 25 percent).[12] Having such a high level of cash might be irrational and inefficient 95 percent of the time, but Intel leadership worried about the 5 percent of the time when catastrophe might devastate the industry or when some other unexpected shock might batter the company.[13] In those rare scenarios, *which inevitably come,* Intel would be able to continue its relentless 20 Mile March, to keep creating, to keep inventing, to keep on its quest to become an enduring great company. Financial theory says that leaders who hoard cash in their companies are irresponsible in their deployment of capital.[14] In a stable, predictable, and safe world, the theory might hold; but the world is not stable, predictable, or safe. And it never will be.

We conducted a systematic analysis of three hundred years of balance sheets from the 10X and comparison companies, and found strong evidence that the 10X cases carried lots of extra oxygen canisters. Compared to the median cash-to-assets ratio for 87,117 companies analyzed in the *Journal of Financial Economics,* the 10X companies carried 3 *to 10 times* the ratio of cash to assets.[15] When it comes to building financial buffers and shock absorbers, the 10X cases were paranoid, neurotic freaks! And it wasn't just an industry effect. When we sliced the data comparing the 10X cases to their comparisons, we found that the 10X cases were more conservative in how they managed their balance sheets than their direct comparisons; 80 percent of the time, the 10X cases carried a higher cash-to-assets ratio and a higher cash-to-liabilities ratio than their comparisons. (See *Research Foundations: Cash and Balance-Sheet-Risk Analysis.*)

We wondered if the 10X cases had adhered to this prudent financial discipline throughout their histories, when they were smaller enterprises *before* they were hugely successful machines spinning out gobs of cash. When we reemployed the same analysis on the first five years after their respective initial public offerings, we found the pattern was already in place, with the 10X cases showing greater financial prudence relative to the comparisons. Intel's conservative cash position in the late 1990s was a continuation of the productive paranoia that its leaders, and their 10X counterparts, adopted in their early years.

Like Breashears and Amundsen, the 10X leaders built buffers and shock absorbers as a habit early on, preparing to absorb the next "Black Swan" event. A Black Swan is a low-probability disruption, an event that almost no one can foresee, a concept popularized by the writer and financier Nassim Nicholas Taleb.[16] Almost no one can predict a *particular* Black Swan before it hits, not even our 10Xers. But it *is* possible to predict that *there will be some* Black Swan, as yet unspecified. Put another way, the probability of any particular Black Swan event might be less than 1 percent, but the probability that *some* Black Swan event will happen is close to 100 percent; it's just that you can't predict what it'll be or when it'll come. This is Taleb's crucial contribution, an insight that any aspiring 10Xer should well learn. 10Xers always prepare for what they cannot possibly predict, stowing away lots of extra oxygen canisters (big margins of safety) and increasing their options *before* they meet the Black Swan—just like David Breashears preparing for Everest.

> 10Xers remain productively paranoid in good times, recognizing that it's what they do *before* the storm comes that matters most. Since it's impossible to consistently predict specific disruptive events, they systematically build buffers and shock absorbers for dealing with unexpected events. They put in place their extra oxygen canisters long *before* they're hit with a storm.

In 1991, Herb Kelleher explained why Southwest Airlines maintained an extremely conservative balance sheet: "As long as we never forget the strengths that enable us to endure and grow in the midst of economic catastrophe; as long as we remember that such economic catastrophes recur with regularity; and as long as we *never* foolishly dissipate our basic strengths through shortsightedness, selfishness, or pettiness, we will continue to endure; we will continue to grow; and we will continue to prosper."[17]

Ten years after he wrote these words, the world watched live, in real time, the horror of September 11, 2001. While the other major airlines cut operations in the immediate aftermath of 9/11, Southwest did not cut a single job or cut a single flight—*not one*—running a full schedule of flights (despite initially flying half-full planes) as soon as the government lifted a national air-travel shutdown. Southwest turned a profit in 2001 (including the fourth quarter of 2001) and was the only major airline to turn a profit in 2002. Southwest opened in new cities, gained market share, and, utterly astonishing, saw its stock price *rise* in the fourth quarter of 2001. At the end of 2002, Southwest achieved a market capitalization greater than all other major U.S. airlines *combined*.[18]

Southwest achieved all this despite what it called "the potentially devastating hammer blow of September 11" because, in its own words from its 2001 annual report, "Our philosophy of managing in good times so as to do well in bad times proved a marvelous prophylactic." On 9/11, Southwest had $1 billion in cash on hand and the highest credit rating in the industry. It also had the lowest cost-per-available-seat-mile, a position secured by thirty years of discipline that never waned during good times. It had a crisis plan in place before 9/11. It had financial-contingency planning tools in place before 9/11. It had nurtured its culture of fierce, caring, and defiant people for thirty years, creating a reciprocal "we'll take care of each other" relationship that proved strong and resilient. If that culture, and those relationships, hadn't been in place before 9/11, Southwest would have suffered like all the other airlines when the terrible event struck.[19]

When Herb Kelleher described how Southwest Airlines responded to 9/11, he showed no personal bravado. He choked on his own tears, unable to finish his sentences, as he tried to describe how Southwest people came together to get the planes in the air as soon as the skies opened, unified in a communal act of defiance.[20] You can attack us, but you cannot beat us; you can try to destroy our freedom, but you'll only make us stronger; you can inflict horror, but you cannot make us terrified. We will fly!

If you come at the world with the practices of building a great enterprise and you apply them with rigor *all the time*—good times and bad, stable times and unstable—you'll have an enterprise that can pull ahead of others *when* turbulent times hit. When a calamitous event clobbers an industry or the overall economy, companies fall into one of three categories: those that pull ahead, those that fall behind, and those that die. The disruption itself does not determine your category. You do.

PRODUCTIVE PARANOIA 2: BOUNDING RISK

We wondered if perhaps 10X companies achieved outsized success simply because they took more risk. Perhaps the 10X cases were just high-risk, high-reward winners, merely lucky that their big risks paid off. But as we got further into the research, we noticed that the 10Xers appeared to lead their companies with a more conservative, risk-averse approach. They constrained growth in the 20 Mile March. They fired bullets before firing cannonballs. They displayed financial prudence, building a cache of extra oxygen canisters. Struck by the accumulating evidence, we undertook a more systematic analysis to ask, "Did the 10X cases take more risk or less risk than the comparison cases?"

To explore this question, we first identified three primary categories of risk relevant to leading an enterprise: (1) *Death Line risk*, (2) *asymmetric risk*, and (3) *uncontrollable risk*. (See *Research Foundations: Risk-Category Analysis*.) Death Line risks are those that could kill or severely damage the enterprise. Asymmetric risks are those for which the potential downside is much bigger than the potential upside. Uncontrollable risks are those that expose the enterprise to forces and events that it has little ability to manage or control. Any particular decision or situation could involve more than one form of risk; the categories of risk are not mutually exclusive.

The Everest story well illustrates these three categories of risk. When Rob Hall decided to abandon the 2:00 p.m. turnaround time to help one of his clients reach the summit, he dramatically increased the risk of being caught in the dark and running out of bottled oxygen—he took an unnecessary Death Line risk. In contrast, David Breashears faced a difficult decision about whether to let a faltering Japanese team member make the final summit bid upon his team's return to the mountain, a "heartbreaking" decision given all the years of effort and training she'd invested. Still, Breashears maintained his margin of safety and didn't let her attempt the summit. Hall's decision to bring enough oxygen canisters for only one summit bid had asymmetric risk. Oxygen canisters are heavy and expensive, but a failed expedition is more expensive, and losing lives is infinitely expensive. Breashears, in contrast, believed that the downside of limited oxygen far outweighed the cost of having an extra cache. Breashears also shunned uncontrollable risk, recognizing that the large number of climbers heading up the mountain on May 8, 1996, could create a situation over which he'd have no control. There could be a dangerous bottleneck at the Hillary Step. Climbers crowding the top could ruin Breashears's summit shot. Breashears and his team could find themselves high on the mountain in a storm, impeded by climbers from the other teams. He chose to avoid these uncontrollable risks by going down on May 8.[21]

Turning to our 10X research data, we conducted an extensive analysis across the history of the 10X and comparison cases, and found that the 10X cases behaved like David Breashears. They took less Death Line risk, less asymmetric risk, and less uncontrollable risk than the comparison cases. The table "Risk Comparison" and the diagram "10X: Less Risk than Comparison" show the results of our analysis.

RISK COMPARISON: 10X COMPANIES VERSUS COMPARISON COMPANIES

Type of Decisions Made	10X Companies	Comparison Companies
Number of Decisions Analyzed	59	55
Decisions Involving Death Line Risk	10% of decisions	36% of decisions
Decisions Involving Asymmetric Risk	15% of decisions	36% of decisions
Decisions Involving Uncontrollable Risk	42% of decisions	73% of decisions
Decisions Classified as Low Risk*	56% of decisions	22% of decisions
Decisions Classified as Medium Risk†	22% of decisions	35% of decisions
Decisions Classified as High Risk‡	22% of decisions	43% of decisions

* Low Risk = no Death Line Risk, no Asymmetric Risk, no Uncontrollable Risk.
† Medium Risk = no Death Line Risk, but one of either Uncontrollable Risk or Asymmetric Risk.
‡ High Risk = Death Line Risk and/or both Asymmetric Risk and Uncontrollable Risk.

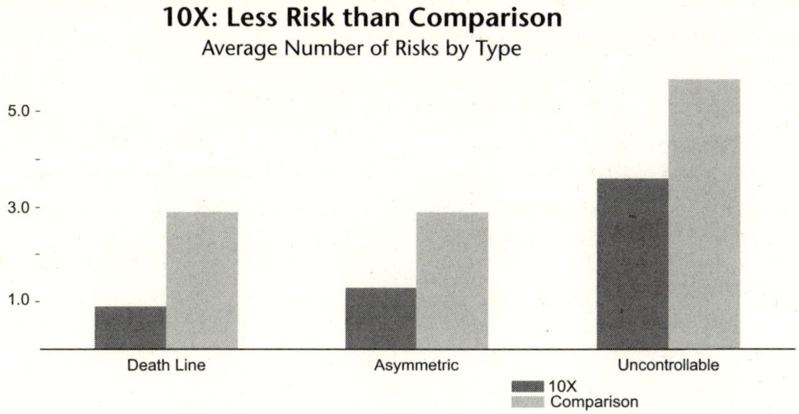

In short, we found that the 10X companies took *less* risk than the comparison cases. Certainly, the 10X leaders took risks, but relative to the comparisons in the same environments, they bounded, managed, and avoided risks. The 10X leaders abhorred Death Line risk, shunned asymmetric risk, and steered away from uncontrollable risk.

After finishing the risk analysis described above, we realized that there was one additional and very important category of risk to consider, *time-based risk*; i.e., when the degree of risk is tied to the pace of events, and the speed of decision and action. If you're facing a tornado roaring across the plains, aimed right for you, your risk profile depends greatly on whether you see the tornado in time, make a decision, and get into a shelter before the tornado reaches you. Given the premise of the study—a turbulent world full of big, fast-moving forces that we can neither predict nor control—perhaps the comparison cases got clobbered by acting too slowly in the face of oncoming risks and disruptions, and the 10X cases reduced their risk through sheer speed.

To test this idea, we identified 115 time-sensitive events in the histories of the 10X and comparison companies. (See *Research Foundations: Speed Analysis.*) We examined the correlation between good and bad outcomes relative to speed of recognition (whether the enterprise recognized the significance of the event early or late), speed of decision, and speed of execution. The table "Speed and Outcomes" summarizes what we learned from this analysis.

SPEED AND OUTCOMES

Behaviors That Correlate with Successful Outcomes	Behaviors That Correlate with Unsuccessful Outcomes
Hypervigilance, constant worry about changes that might signal danger; early recognition of threat.	Arrogance; minimization or ignorance of the potential significance of changes; late recognition of threat.
Adjustment of decision speed to the pace of events, whether fast or slow—"go slow when you can, fast when you must."	Failure to adjust decision speed to the pace of events, deciding too slowly or too fast depending on the situation.
Deliberate, fact-driven decisions; highly disciplined thought, no matter how fast.	Reactive, impulsive decisions, lacking fanatic discipline and strategic rigor.
Focus on superb execution once decisions are made; intensity increased as needed to meet time demands without compromising excellence.	Compromise in excellence of execution for the sake of speed; failure to increase intensity to ensure superb execution when moving fast.

As the table demonstrates, our analysis yielded a much more nuanced perspective than "always go faster." We concluded that recognizing a change or threat *early*, and then taking the time available—whether that be short or long—to make a rigorous and deliberate decision yields better outcomes than just making a bunch of quick decisions. The key question turns out not to be, "Should we be fast to act or slow?" but *"How much time before our risk profile changes?"*

Recall Andy Grove's response to his cancer diagnosis that we discussed in Chapter 2. He didn't jump right to action. He considered his time frame and recognized that his risk profile wouldn't change significantly in a matter of weeks. Months or years, yes, but not weeks. He then used that time to rigorously develop a plan of attack, considering all the various possibilities and creating his own data charts. Grove was anything but complacent about his cancer, but he didn't make a quick, reactive decision. Grove believed that jumping into the operating room without carefully considering his situation and the options would *increase* his risk.[22]

> Sometimes acting too fast increases risk. Sometimes acting too slow increases risk. The critical question is, *"How much time before your risk profile changes?"* Do you have seconds? Minutes? Hours? Days? Weeks? Months? Years? Decades? The primary difficulty lies not in answering the question but in having the presence of mind to *ask* the question.

The 10X teams tended to take their time, to let events unfold, when the risk profile was changing slowly; yet equally, they prepared to act blindingly fast in the event that the risk profile began to change rapidly. Prior to the mid-1990s, Stryker kept a vigilant eye on a storm brewing far off in the distance, noting in its 1989 annual report that the United States would become competitively disadvantaged if healthcare costs were to rise to more than 15 percent of GNP; this, in turn, could lead to a backlash on costs and drive down prices on Stryker's medical devices. Stryker squirreled away a whole bunch of oxygen canisters (cash on the balance sheet) to handle whatever form the disruption might take. (See diagram "Stryker: Preparing for a Storm.") Still, John Brown did *not* act early; he let the situation unfold, *prepared* to act fast *when* the time came.[23]

Then in the late 1990s, Stryker's risk profile began to change rapidly when hospital buying groups emerged to concentrate their buying

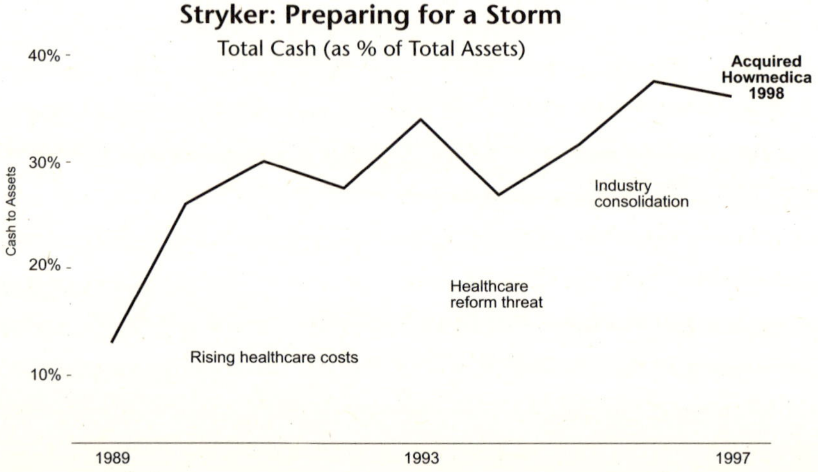

power. These groups preferred to deal with a few large market leaders, and in response, the industry began to consolidate in a series of rapid-fire acquisitions. Medical-device companies faced a stark choice: become one of the few largest players, with economies of scale, or be largely shut out of the game. And *that's* when Stryker swooped in, bought Howmedica, and ensured itself one of the top three seats.[24]

SPEED AND OUTCOMES: STRYKER, HEALTHCARE COSTS, AND INDUSTRY DISRUPTION[25]

Behaviors That Correlate with Successful Outcomes	Stryker's Behaviors
Hypervigilance, constant worry about changes that might signal danger; early recognition of threat.	In the 1980s, Stryker explicitly identified rising healthcare costs as a concern and worried about industry disruptions that might result.
Adjustment of decision speed to the pace of events, whether fast or slow—"go slow when you can, fast when you must."	In the 1980s and early 1990s, Stryker took no dramatic action yet considered options and built large cash reserves.

Deliberate, fact-driven decisions; highly disciplined thought, no matter how fast.	In the late 1990s, buying groups drove the industry into rapid consolidation; Stryker made the disciplined decision to buy Howmedica.
Focus on superb execution once decisions are made; intensity increased as needed to meet time demands without compromising excellence.	From 1998 to 1999, Stryker team members worked nearly nonstop to successfully integrate Howmedica.

As a productive paranoid, you want to be cognizant of lurking dangers and vigilant about possible disruptions, but this is very different from taking quick, immediate action because you want the anxiety and uncertainty to go away. In our executive laboratory, we've noticed that some leaders from emerging markets maintain a very calm stance in the face of uncertainty, including a willingness to let time pass when the risk profile remains stable. During the 2008–9 financial crisis, we worked directly with some of the most successful business leaders from emerging markets, and we noticed their calm and considered countenance in the face of swirling tumult. One of the most successful self-made business leaders from Latin America who achieved his success in a brutally uncertain environment described his ability to pause, "Sure, it's human nature to want to make the uncertainty go away. But that desire can lead you to decide quickly, sometimes too quickly. Where I come from, you soon realize that uncertainty will *never* go away, no matter what decisions we make or actions we take. So, if we have time to let the situation unfold, giving us more clarity before we act, we take that time. Of course, when the time comes, you need to be ready to act."[26] One of the most dangerous false beliefs is that faster is always better, that the fast always beat the slow, that you are either the quick or the dead. Sometimes the quick *are* the dead.

PRODUCTIVE PARANOIA 3: ZOOM OUT, *THEN* ZOOM IN

In a famous experiment, researchers Daniel J. Simons and Christopher F. Chabris asked subjects to watch a videotape of people passing a basketball back and forth, and to count the number of passes; partway through the video, a person in a gorilla suit unexpectedly walks right into the middle of the action, thumps its chest, and walks off the court. Focused on their counting task, only 50 percent of the subjects even noticed the gorilla.[27]

We spend most of our lives dealing with the plans and activities right in front of us, ticking tasks off our lists, clicking past mile markers on our big projects, responding to the incessant demands on our time. And we can easily miss the gorilla right in front of us. 10X leaders, however, don't miss the gorilla, especially if the gorilla poses a dangerous threat. David Breashears was utterly focused on getting his IMAX camera to the summit of Everest, yet when he looked down the mountain on May 8, 1996, the swarm of humanity heading his way, he saw a huge gorilla.

> We adopted the terms *zoom out* and *zoom in* to capture an essential manifestation of productive paranoia, a dual-lens capability. 10X leaders remain obsessively focused on their objectives *and* hypervigilant about changes in their environment; they push for perfect execution *and* adjust to changing conditions; they count the passes *and* see the gorilla.

In practice, it works like this:

Zoom Out

Sense a change in conditions

Assess the time frame: How much time before the risk profile changes?

Assess with rigor: Do the new conditions call for disrupting plans? If so, how?

Then

Zoom In

Focus on supreme execution of plans and objectives

Notice that the question "How much time before the risk profile changes?" is part of the *zoom out*. As we discussed earlier in the chapter, 10Xers took the time available to *zoom out* and formulate a considered response. Of course, sometimes the 10X cases had to act fast, when the risk profile was changing rapidly, when the gorilla was already close and charging fast. Even so, they avoided panicky, reactive decisions; they remained deliberate and clearheaded, responding fast *enough*.

On December 4, 1979, a special task force of six Intel managers and external marketing guru Regis McKenna set aside their lives for three full days of intense discussions, sparked by an eight-page telex from field engineer Don Buckhout, who'd written an "incisive and desperate" analysis of Intel's weakening position with its 8086 microprocessor relative to the Motorola 68000.[28] Of particular concern, Motorola had begun to pull ahead of Intel in competition for important "design wins," convincing customers to design the Motorola 68000 into their product lines. It was a terrifying trend; if Motorola gained a dominant share of design wins, it could entrench itself as a standard, becoming increasingly difficult to unseat. As Intel manager William H. Davidow reflected in his book, *Marketing in High Technology*, "Intel was headed for obscurity."[29]

The team *zoomed out*. Why is Motorola winning? How important is this? How can we counter? The team developed a five-point competitive positioning strategy and a schedule, focusing on Intel's distinctive capability, "Intel Delivers," and its capacity to provide an entire family of chips, generation upon generation, giving customers comfort. The resulting document was smart and strategic, reflecting deep insight into Intel's strengths and an understanding of what customers really worried about. Based on a very systematic analysis, the team developed a plan of counterattack dubbed Operation CRUSH.[30]

Then Intel *zoomed in*. The task force finished its work on Friday, less than a week after it convened, and Intel approved the plan and allocated a multimillion-dollar budget the following Tuesday. Within the week, more than a hundred CRUSH team members, sporting buttons with the bold, orange letters C-R-U-S-H, met at the San Jose Hyatt. From there, they fanned out across the globe to garner two *thousand* design wins for Intel within a year. Intel was on a self-described crusade, turning the tide, and getting its two thousand design victories, including a really big one for IBM's future PC.[31]

> Despite being in a fast-moving, perilous, competitive situation, the Intel team took a very deliberate approach, formulating a smart and rigorous strategy. Intel initiated Operation CRUSH in just seven days yet did so with fiercely disciplined thought. When facing fast-moving threats, 10X teams neither freeze up nor immediately react; they think first, even when they need to think fast.

Intel made a mistake in not recognizing the Motorola threat earlier (even 10X companies do not have a perfect record), which forced it into such a crash program. Yet once it recognized the threat, it didn't make its situation worse via panicky, thoughtless reaction. 10X enterprises at their best respond to empirical evidence rather than hype or scaremongering, and stick with proven principles and strategies in the face of frightening events. A fast-moving threat does not call for abandoning disciplined thought and disciplined action.

By early 1987, George Rathmann had convinced the Amgen board to fire a cannonball on its breakthrough product, EPO. Recognizing the moment—the science is done, the trials are done, we have the product ready, the clock is ticking, we've got to go *now*!—the Amgen FDA-application team turned themselves into the "Simi Valley Hostages."

At first, they worked at the office but soon decided that they needed to block out all distractions, recognizing that nothing was more important at this moment than the FDA filing, pushing everything else into the "it can wait" pile, moving copiers and working files into motel rooms at the Posada Royale Quality Inn in Simi Valley, cutting off normal life, embracing a brutal, non-stop schedule, smiling as their friends and colleagues hung yellow ribbons in their honor. They'd work in the morning, take a brief lunch break, work until 6:00 p.m., take a short dinner break, work into the night, then repeat again, day after day, week after week. Finally, 93 days later, they loaded the 19,578-page document into a rented truck, drove it to the airport, and shipped it off to the FDA. A large bedsheet adorned with yellow ribbons was hung outside Amgen's headquarters proclaiming, "The Simi hostages are *free*!"[32]

The Simi Valley Hostages had a lot of catching up to do on their lives. If you don't get your desk cleaned off for 93 days, or your garage repainted, your marathon run, your golf game in, your expense reports completed, your phone calls returned, your mail answered, your vacation achieved, your new house bought, your newspapers read, or pretty much anything else that can wait until later, what does it matter, compared to missing the chance to secure EPO with the FDA before another competitor?

> The Simi Valley Hostages understood that they were in a race to be first, but they didn't sacrifice their detailed, methodical approach for the sake of speed. By increasing their intensity to extreme levels for a time—*nothing else matters until we get this done, and done right!*—they went fast enough to win.

In contrast, consider how Genentech failed to execute as well in a comparable moment and how that failure contributed to Genentech losing its independence. On Friday afternoon, May 29, 1987, four hundred people gathered in the FDA auditorium in Bethesda, Maryland, to witness Genentech's presentation to an FDA advisory panel on its new drug t-PA (also known as Activase). No drug up to that point in the history of biotechnology had generated anything like the excitement around t-PA, a wonder drug designed to dissolve clots in heart-attack victims. Genentech's stock price, trading at a hundred times earnings, reflected Genentech's salesmanship in convincing people that the t-PA cannonball would smash directly into its target—hype that'd leave the stock vulnerable in the event that t-PA encountered snags with the FDA.[33]

Around dinnertime, after five hours of presentations and discussions, the committee chair finally asked for a vote. The audience gasped when it heard the count.[34] Genentech hadn't convinced the committee that t-PA prolonged life, and the committee recommended that t-PA should be sent back for further study.[35] Ironically, Genentech actually *had* access to most of the information needed to convince the FDA, but it didn't have all the necessary data readily available and prepared in an unassailable way for whatever concerns and questions might have come from the committee on the day of the meeting.[36]

Genentech's founder, Robert Swanson, called the committee's decision a mistake, and to be fair, Genentech did return later that year and obtained a positive decision.[37] Still, those six months mattered, with at least ten companies racing to create t-PA–related drugs, and those competitors gained ground while Genentech retreated to reassemble its data for the FDA.[38] The t-PA setback helped to puncture Genentech's high-flying stock price, which fell more than 60 percent behind the general stock market in the subsequent two years, raising the cost of equity capital (which Genentech needed to invest heavily in R&D) and leading the company to sell a controlling stake to Roche.[39]

NOT ALL TIME IN LIFE IS EQUAL

We close this chapter with a twist to the Amundsen story that highlights the importance of being able to *zoom out,* then *zoom in.* It turns out that Amundsen hadn't planned to go to the South Pole in 1911; he'd planned to go to the North Pole.

That's right, the *North* Pole!

He'd raised money to go to the North Pole, assembled a team for the North Pole, gained access to the ship *Fram* for a trip to the North Pole, and mapped a full plan for the North Pole.[40]

So then, how did he end up at the opposite end of the Earth, at the South Pole?

While making his preparations for the North, Amundsen received crushing news. The North Pole had fallen. First Cook, then Peary, had reportedly reached 90 degrees North. So, Amundsen decided to redirect his expedition and channeled his energies into preparing for a new destination, the South Pole. He kept his decision secret, even from his crew, during the months while he prepared until he set sail. On September 9, 1910, at the port of Madeira, Portugal, Amundsen raised anchor three hours ahead of schedule, catching his crew off guard. He assembled his men on deck and calmly told them that they weren't going to the North Pole after all, that the expedition would veer to the South Pole instead. Earlier in the day, the crew had nothing but the North Pole on their minds; by 10:00 p.m. they were already heading toward the South Pole, fully committed to the new adventure, the North Pole fading from their dreams.[41]

We've portrayed Amundsen as anything but impulsive, the consummate detail-oriented, super-prepared, monomaniacal, disciplined fanatic. Yet with the North Pole gone and the South in Scott's line of sight, he pivoted dramatically, changing direction from north to south. If Amundsen had said, "Well, my plan is to go north, so that's what I'm going to do," if he refused to reorient his focus, he would not have led his team to a 10X achievement. Upon learning that the North Pole had fallen, he *zoomed out* to consider the changed conditions; then he *zoomed in* to execute a new plan to go south.

10Xers distinguish themselves by an ability to recognize defining moments that call for disrupting their plans, changing the focus of their intensity, and/or rearranging their agenda, because of opportunity or peril, or both. When the defining moment comes, they have the buffers already in place, lots of extra oxygen canisters, giving them options and the flexibility to adjust. They have huge margins of safety, precisely because they've bounded their risks, exercising prudence all the way along, avoiding Death Line risk, shunning asymmetric risk, and minimizing uncontrollable risk. They sense change, *zooming out* to ask, "How much time before the risk profile changes?" They make rigorous rather than reactive decisions. Then they *zoom in*, obsessively focusing on superb execution in the defining moment, never compromising excellence for speed.

Not all time in life is equal. Life serves up some moments that count much more than other moments. The year 1911 was an unequal time for Amundsen, and he made the most of it. May 1996 on Everest was an unequal time for David Breashears, and he executed brilliantly when the time came. September 11th was an unequal time for the airline industry, and Southwest came through with the most inspired and defiant performance. We will all face moments when the quality of our performance matters much more than other moments, moments that we can seize or squander. 10Xers prepare for those moments, recognize those moments, grab those moments, upend their lives in those moments, and deliver their best in those moments. They respond to unequal times with unequal intensity, when it matters most.

LEADING ABOVE THE DEATH LINE

KEY POINTS

▶ This chapter explores three key dimensions of productive paranoia:

1. Build cash reserves and buffers—*oxygen canisters*—to prepare for unexpected events and bad luck *before* they happen.
2. Bound *risk—Death Line risk, asymmetric risk,* and *uncontrollable risk*—and manage time-based risk.
3. *Zoom out,* then *zoom in,* remaining hypervigilant to sense changing conditions and respond effectively.

▶ 10Xers understand that they cannot reliably and consistently predict future events, so they prepare obsessively—ahead of time, all the time—for what they cannot possibly predict. They assume that a series of bad events can wallop them in quick succession, unexpectedly and at any time.

▶ It's what you do before the storm hits—the decisions and disciplines and buffers and shock absorbers already in place—that matters most in determining whether your enterprise pulls ahead, falls behind or dies when the storm hits.

▶ 10Xers build buffers and shock absorbers far beyond the norm of what others do. The 10X companies we studied carried 3 *to 10 times* the ratio of cash to assets relative to the median of what most companies carry and maintained more conservative balance sheets than the comparison companies throughout their histories, even when they were small enterprises.

▶ 10X cases are extremely prudent in how they approach and manage risk, paying special attention to three categories of risk:

1. Death Line risk (which can kill or severely damage the enterprise)
2. Asymmetric risk (in which the downside dwarfs the upside)
3. Uncontrollable risk (which cannot be controlled or managed)

▶ 10Xers *zoom out*, then *zoom in*. They focus on their objectives *and* sense changes in their environment; they push for perfect execution *and* adjust to changing conditions. When they sense danger, they immediately *zoom out* to consider how quickly a threat is approaching and whether it calls for a change in plans. Then they *zoom in*, refocusing their energies into executing objectives.

▶ Rapid change does not call for abandoning disciplined thought and disciplined action. Rather, it calls for upping the intensity to *zoom out* for fast yet rigorous decision making and *zoom in* for fast yet superb execution.

UNEXPECTED FINDINGS

▶ The 10X cases took less risk than the comparison cases yet produced vastly superior results.

▶ Contrary to the image of brazen, self-confident, risk-taking entrepreneurs who see only upside potential, 10X leaders exercise productive paranoia, obsessing about what can go wrong. They ask questions like: What is the worst-case scenario? What are the consequences of the worst-case scenario? Do we have contingencies in place for the worst-case scenario? What's the upside and what's the downside of this decision? What's the likelihood of the upside and the downside? What's out of our control? How can we minimize our exposure to forces we can't control? What if? What if? What if?

▶ The 10X cases didn't have a greater bias for speed than the comparison companies. Taking the time available before the risk profile changes, whether short or long, to make a rigorous and deliberate decision produces a better outcome than rushing a decision.

ONE KEY QUESTION

▶ Regarding the biggest threats and dangers facing your enterprise, *how much time before the risk profile changes?*

6

SMaC

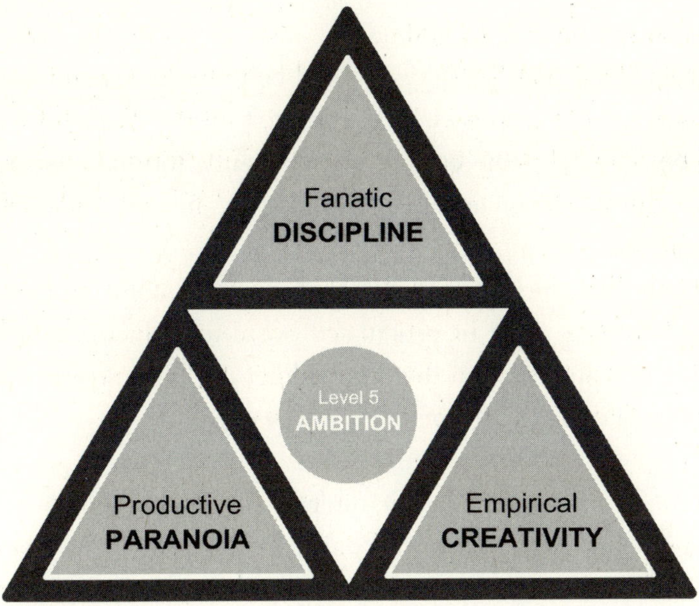

"Most men die of their remedies, and not of their illnesses."

—Molière[1]

In early 1979, Howard Putnam, then CEO of Southwest Airlines, wrestled with a question: does the sweeping disruption of deregulation call for a revolution in how we run our company? The 1978 Airline Deregulation Act would unleash competition, throw carriers into pitched battles for market share, ignite price wars, force airlines to cut costs, and lead to bankruptcies.

Putnam considered: Does deregulation undermine our low-cost model? Does deregulation threaten our high-spirit, employee-focused culture? Does deregulation erode the competitive value of rapid gate turns or destroy the viability of our point-to-point system? Does radical change in our environment call for inflicting radical change upon ourselves?[2]

His answers: no, no, no, and no.

He concluded that Southwest should continue to expand based on "the 'cookie-cutter' approach." He conjured up the image of a recipe used repeatedly to create batches of consistently formed cookies. "Do the same thing that you are already doing well," he said, and do it "over and over again."

Not only that, he specified the cookie recipe, point by point. Reproduced below is what he articulated (we're reproducing it *verbatim*, excluding one abbreviation that we couldn't decipher, so that you can see how he laid out the recipe in his own words):[3]

1. Remain a short-haul carrier, under two-hour segments.
2. Utilize the 737 as our primary aircraft for ten to twelve years.
3. Continued high aircraft utilization and quick turns, ten minutes in most cases.
4. The passenger is our #1 product. Do not carry air freight or mail, only small packages which have high profitability and low handling costs.
5. Continued low fares and high frequency of service.
6. Stay out of food services.
7. No interlining . . . costs in ticketing, tariffs and computers and our unique airports do not lend themselves to interlining.
8. Retain Texas as our #1 priority and only go interstate if high-density short-haul markets are available to us.
9. Keep the family and people feeling in our service and a fun atmosphere aloft. We're proud of our employees.

10. Keep it simple. Continue cash-register tickets, ten-minute cancellation of reservations at the gate in order to clear standbys, simplified computer system, free drinks in Executive service, free coffee and donuts in the boarding area, no seat selection on board, tape-recorded passenger manifest, bring airplanes and crews home to Dallas each night, only one domicile and maintenance facility.

Putnam didn't issue some bland, generic "Southwest Airlines will be a leading low-cost airline" vacuous statement. He specified two-hour segments. He specified 737s. He specified 10-minute turns. He specified no air freight or mail. He specified no food service. He specified no interlining. He specified no seat selection. He specified cash-register receipts. Putnam's 10 points are easy to grasp, articulate, follow, and understand what to do and what not to do. Putnam laid out a clear, simple, and concrete framework for decisions and action.

Putnam's 10 points reflect insight, based upon empirical validation about *what works.* Take the idea of only 737s. Why would only 737s make sense? All your pilots can fly all your jets, allowing for immense scheduling flexibility. You need only one set of parts, one set of training manuals, one set of maintenance procedures, one set of flight simulators, one type of jet way, one procedure for boarding.

But the truly amazing thing about Putnam's list is its consistency over time. In total, the elements on the Putnam list *changed only about 20 percent in a quarter of a century.* Stop to think about that for a moment: only a 20 percent change, despite a series of disruptive events from fuel shocks to air-traffic-control strikes, massive industry mergers, the rise of the hub-and-spoke model, recessions, interest-rate spikes, the Internet, and 9/11. Yet while stunningly consistent, the recipe also evolved—never through wholesale revolution but in careful steps. Southwest did eventually add flights longer than two hours, embraced Internet booking, and interlined with Icelandair.[4] If Southwest had become rigid, close-minded, uncurious, never amending Putnam's points as needed, it would not have become a 10X case.

Still, what most stands out is how much of the list Southwest kept intact.

SMaC RECIPE

Howard Putnam's 10 points form a SMaC recipe. A SMaC recipe is a set of durable operating practices that create a replicable and consistent success formula. The word "SMaC" stands for Specific, Methodical, and Consistent. You can use the term "SMaC" as a descriptor in any number of ways: as an adjective ("Let's build a SMaC system"), as a noun ("SMaC lowers risk"), and as a verb ("Let's SMaC this project"). A solid SMaC recipe is the operating code for turning strategic concepts into reality, a set of practices more enduring than mere tactics. Tactics change from situation to situation, whereas SMaC practices can last for decades and apply across a wide range of circumstances.

> We on the research team used to believe in an inevitable trade-off between specificity and durability: if you want to have durable precepts to live by, they need to be more general, like core values or high-level strategy; but if you want specific practices, they need to change frequently as conditions change, like tactics. Yet it is possible to develop practices that are both specific *and* durable—SMaC practices.

A SMaC practice is not the same as a strategy, culture, core values, purpose, or tactics.

Is "Fly only 737s" a core value? No.

Is "Fly only 737s" a core purpose, a reason for being? No.

Is "Fly only 737s" a high-level strategy? No.

Is "Fly only 737s" a culture? No.

Is "Fly only 737s" a tactic to be changed frequently, from situation to situation? No. More than thirty years after Putnam laid out his 10 points, Southwest *still* flew only 737s.[5]

A SMaC recipe also includes practices "*not* to do." Putnam's list has clear not-to-do points—don't interline, serve food, offer first-class seats, or carry air freight. Putnam grasped that adding any of these services would complicate the process of getting planes turned around fast. All the 10X companies' SMaC recipes contained things not to do. Do *not* use loss reserves to manage earnings (Progressive). Do *not* wait to develop perfect software to enter the market; get good enough to launch and then improve (Microsoft). Do *not* be the first with new innovations but also *not* the last; stay one fad behind (Stryker). Do *not* cut R&D during industry recessions (Intel). Do *not* hype; better to make people angry by underestimating your next success than by overestimating (Amgen). Do *not* grant stock options to the CEO but only to employees (Biomet).[6]

The clarity and specificity of a SMaC recipe helps people keep their bearings and sustain high performance when in extreme conditions. Think back to David Breashears on Everest. Over the years leading up to the IMAX project, he developed a SMaC recipe for filmmaking in the high mountains. He went to a 50-degree-below-zero-F freezer in Toronto to develop specific protocols for handling the IMAX camera in extreme cold, assessing how the batteries would perform and practicing loading the 65mm film with bare hands. (Even on the top of Everest, he had to load the camera with bare hands to minimize any chance of malfunction.) He created an "Idiot Check" list for working and moving the camera in extreme conditions and unusual situations. He systematically developed a supply list that eliminated any weight that didn't directly contribute to the IMAX project or to safety. He then refined all his methods on a 160-mile, 28-day trek in Nepal the year before the Everest ascent. By the time he and his team were filming on Everest, they knew exactly what to do and *precisely* how to do it. On May 23, 1996, Breashears and his team stood on top of Everest with the IMAX camera. One mistake—a dropped piece of camera equipment, a malfunction, a bungled film feed—could wipe out years of effort and millions of dollars expended. "We worked slowly and methodically just as we had for the past sixty days," explained Breashears of the crucial moment. "Bare-handed, I threaded the film again. Then, at the apex of the world, Robert and I went over our camera checklist one last time." SMaC![7]

INGREDIENTS IN DAVID BREASHEARS'S SMaC RECIPE[8]

1. Create a binder with individual tabs for all facets of the expedition, including backup plans (and sometimes even backup plans to the backup plans) for everything that can plausibly go wrong.

2. Perform "Idiot Check" every time you move locations—360-degree spin to make sure you haven't left anything behind.

3. Thread the camera with bare hands, no matter how cold, to ensure a perfect shot every time.

4. Be able to assemble the camera, mount it on the tripod, load and thread film, aim, and shoot in five minutes flat.

5. Test equipment in real conditions, sub-zero-freezer and simulation trips before the actual expedition.

6. Always optimize weight and functionality. Carry the least amount of mass without sacrificing function/safety.

7. In selecting teammates, choose people to get stranded with.

8. Always bring backups for critical gear and supplies: extra oxygen, extra crampons, extra mittens, and extra supplies. Be prepared to stay longer than planned.

9. Never let a weak member attempt to summit. "A team is only as strong as its weakest member."

10. Have two separate teams, climbers and filmmakers, that work well together on the mountain.

In a world full of big, fast-moving forces and unrelenting uncertainty, 10Xers accept with stoic equanimity what they cannot control, yet they exert extreme control when they can. One of the most crucial ways they exert control in an out-of-control world is by being incredibly SMaC. The more unforgiving your world, the more SMaC you need to be. A SMaC recipe forces order amidst chaos. It imposes consistency when you're slammed by disruption. Operating in a turbulent world without a SMaC recipe is like being lost in the wilderness in the middle of a storm without a compass.

Now, you might be thinking, "OK, the primary finding here is to have a SMaC recipe." But in fact, the existence of a recipe per se did *not* systematically distinguish the 10X companies from the comparison companies. Rather, the principal finding is how the 10X companies *adhered* to their recipes with fanatic discipline to a far greater degree than the comparisons, and how they carefully *amended* their recipes with empirical creativity and productive paranoia.

ADHERING TO THE SMAC RECIPE WITH FANATIC DISCIPLINE

The 10X companies kept any given recipe ingredient in the mix for more than twenty years on average (with a range from eight to forty-plus years)—durable indeed! The table "Progressive Insurance SMaC Recipe" illustrates the durability and consistency of a 10X SMaC recipe.

Progressive Insurance SMaC Recipe[9]	Durability and Consistency
1. Concentrate on non-standard auto insurance, insuring high-risk drivers whom major insurance companies would likely turn away.	30+ years Changed in the 1990s
2. Price to achieve 96% combined ratio. Price for profitability, never for growth; never lower underwriting standards or pricing discipline to increase market share. There's no excuse for failing to deliver an underwriting profit, not regulatory problems, not competitive difficulties, not natural disaster, nothing.	30+ years No change as of 2002

(continued on next page)

Progressive Insurance SMaC Recipe[9]	Durability and Consistency
3. Price for each individual customer, based on every available piece of information on that person's life that might impact driving risk (such as zip code, age, marital status, driving record, vehicle make and year, size of engine), even if that means thousands of different pricing premiums.	30+ years No change as of 2002
4. Exit any state where regulation makes profitable pricing with superb claims service impossible.	20+ years No change as of 2002
5. Focus on speed in claims adjustment; speed results in better service and lower costs.	25+ years No change as of 2002
6. Have at least one new business or service experiment under way yet keep any new business to less than 5% of total revenues until it demonstrates sustained profitability.	30+ years No change as of 2002
7. Deliver profits primarily from underwriting, not investing.	30+ years No change as of 2002
8. Never use loss reserves to manage earnings.	30+ years No change as of 2002
9. Employ independent agents as our sales force; do a small amount of business with a large number of agents rather than a large amount of business with a small number of agents.	30+ years Changed in the 1990s

We found a fascinating contrast in the comparison cases: most of the comparisons displayed some version of a SMaC recipe during their best years of performance (only one comparison company, Kirschner, never had one) but *the comparisons changed their recipes to a much greater degree* than the 10X cases over time. When we analyzed 117 recipe elements across the 10X and comparison cases, we found that the comparisons changed *four times more* than the 10X cases. (See *Research Foundations: SMaC-Recipe Analysis.*) The table "Change in Ingredients in SMaC Recipes" shows how much the 10X cases and comparison cases changed their recipes over their respective eras of analysis.

CHANGE IN INGREDIENTS IN SMaC RECIPES DURING THEIR RESPECTIVE ERAS OF ANALYSIS

10X Company		Comparison Company	
Amgen	10%	Genentech	60%
Biomet	10%	Kirschner	N/A
Intel	20%	AMD	65%
Microsoft	15%	Apple	60%
Progressive	20%	Safeco	70%
Southwest Airlines	20%	PSA	70%
Stryker	10%	USSC	55%

Now, you might be thinking, "But wait a minute! Perhaps the comparison cases had truly inferior operating models, and they changed more because they hadn't yet found a great one." But think back on PSA. Recall from Chapter 4 how Southwest Airlines began as a copy of PSA, right down to the operating manuals. So, here we have two airlines both facing deregulation, both facing a disruptive environment, both with fabulous core markets, both with *nearly identical recipes*, and yet only Southwest endured as a great company in the two decades after deregulation.

PSA reacted to deregulation by deciding it needed to become more like . . . United Airlines. Here, in an amazing twist of irony, we have PSA moving away from its proven recipe just as Southwest began to build momentum in Texas. Using the same proven recipe, and having *invented* it, PSA should have become the most successful airline in history, yet it sold out to US Air. "Life is tough for an independent airline at the best of times," said PSA's president, ending the company's independent life with a whimper. "We could have gone it alone, but . . . it made more sense for us to accept US Air's very reasonable offer."[10]

Analysts and the media began chanting that Southwest, the genetic twin of PSA's original concept, also needed to change its formula, that Putnam's simple list needed major revision, otherwise it might go down like PSA. "A growing chorus of critics says the 56-year-old Kelleher needs to rethink his keep-it-simple strategy," wrote *Business Week* in 1987. The *Wall Street Transcript* quoted analysts saying that Southwest could no longer be viewed as a growth company, its model running out of opportunity. Herb Kelleher, by then CEO, responded to this pressure to revolutionize the airline much as General McAuliffe responded to the German surrender ultimatum at the Battle of the Bulge: "Nuts!" Kelleher understood why each ingredient in Putnam's list worked, and he understood that the Southwest model would still apply in an increasingly competitive airline industry. He kept most of the recipe intact. Southwest Airlines, of course, went on to become one of the most admired companies in the world, while PSA became irrelevant, then forgotten. The PSA spirit endured, but deep in the heart of Texas.[11]

Airline Deregulation: Different Responses by Different Airlines

Southwest Airlines vs. PSA

Ratio of Cumulative Stock Returns to General Market

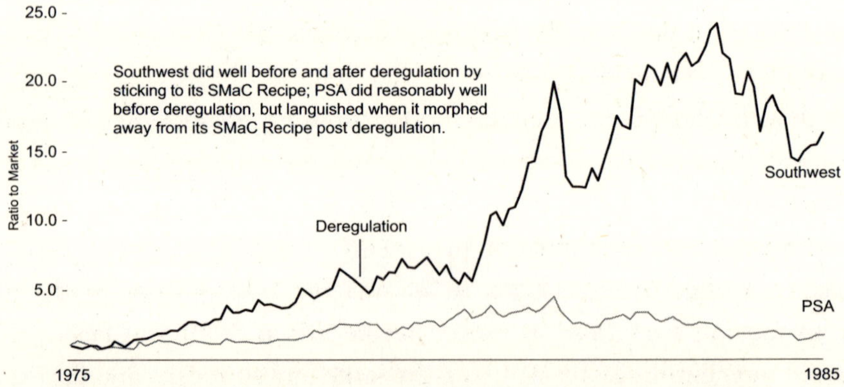

Notes:
1. Each company's ratio to market was calculated from December 31, 1974 to December 31, 1984.
2. Source for all stock return calculations in this work: ©200601 CRSP®, Center for Research in Security Prices. Booth School of Business, The University of Chicago. Used with permission. All rights reserved. www.crsp.chicagobooth.edu.

> Conventional wisdom says that change is hard. But if change is so difficult, why do we see more evidence of radical change in the less successful comparison cases? Because change is *not* the most difficult part. Far more difficult than implementing change is figuring out what works, understanding why it works, grasping when to change, and knowing when not to.

The fall and rise of Apple illustrates the danger of straying from a recipe and the value of restoring it. By the mid-1990s, Apple had fallen far from its glorious early days, when it had brought forth the Apple II and then the Mac, "the computer for the rest of us." Beset by chronic inconsistency, it had a revolving door at the top; John Sculley ousted Steve Jobs in 1985, Michael Spindler replaced John Sculley in 1993, Gil Amelio replaced Michael Spindler in 1996. It also lurched back and forth in its positioning: computers for the rest of us, then computers for business, then premium-priced BMWs of computers, then low-cost machines in a high-market-share strategy, then back again to premium machines. Apple's stock returns fell behind the general stock market, in stark contrast to Microsoft's upward march. (See diagram "1985–1997: Microsoft Soared, Apple Faltered.") Microsoft during this time showed unwavering consistency—consistency in leadership, consistency in purpose, consistency in strategy, consistency in recipe. By 1993, Apple had fallen so far behind that a technology conference featured a panel of venture capitalists and computer-industry experts debating the hot topic, "Will Apple Computer Survive?"[12] Apple eventually began serious talks with companies like Sun Microsystems about selling itself, itching to fire a bullet in the head of its own independence. It looked like Apple's quest to be a great company would die an inglorious death.[13]

1985–1997: Microsoft Soared, Apple Faltered

Ratio of Cumulative Stock Returns to General Market

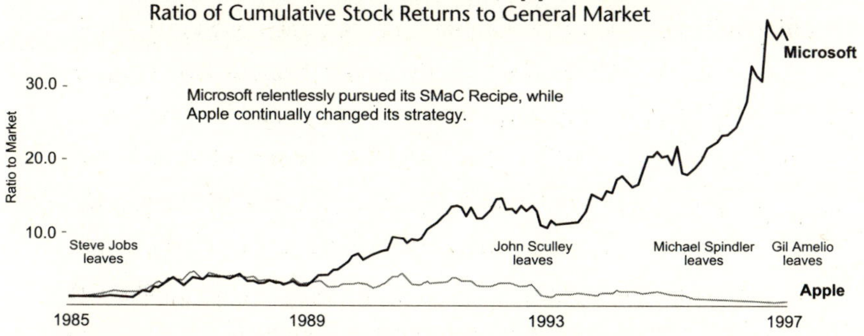

Notes:

1. Each company's ratio to market was calculated from August 31, 1985 to August 31, 1997. Also, because CRSP data is not available for Microsoft from August 1985 to March 1986, general stock market returns were used as a proxy for company returns during this time period.
2. Source for all stock return calculations in this work: ©200601 CRSP®, Center for Research in Security Prices. Booth School of Business, The University of Chicago. Used with permission. All rights reserved. www.crsp.chicagobooth.edu.

Fortunately, the story turned out differently, with a turnaround beginning in 1997. And here's the really interesting part: Steve Jobs didn't so much revolutionize the company as he *returned* it to the principles he'd used to launch the company from garage to greatness two decades earlier. "The great thing is that Apple's DNA hasn't changed," said Jobs in 2005.[14] And not just its larger purpose but also many of its recipe ingredients. For example: allow no one else to clone our products; design our products so they work seamlessly together; make design friendly and elegant; obsess about secrecy and then do big launches to capture pent-up excitement; don't enter any business where we don't control the primary technology; design for and market to individuals, not businesses. All of these practices were in place during Apple's early days and were then brought back to life during Apple's rebirth two decades later. Apple fell behind during its dark days *not* because its original recipe no longer worked, but *because it lacked the discipline to adhere to its original recipe.* Steve Jobs's genius notwithstanding, Apple roared back because it returned, this time with fanatic discipline, to the essence of its original recipe. As John Sculley commented in a 2010 interview, reflecting upon the resurgence of Apple under the leadership of the very man he'd ousted 25 years earlier, "The same principles Steve is so rigorous about now are the identical ones he was using then."[15]

> When faced with declining results, 10Xers do not first assume that their principles and methods have become obsolete. Rather, they first consider whether the enterprise has perhaps strayed from its recipe, or has forgone discipline and rigor in adhering to the recipe. If so, they see the remedy in reconnecting with the underlying insights behind the recipe and reigniting passion for adhering to it. They ask, "Is our recipe no longer working because we've lost discipline? Or is it no longer working because our circumstances have fundamentally changed?"

John Wooden, the great UCLA basketball coach who produced 10 NCAA championship teams in 12 years during the 1960s and 1970s, perfectly exemplified the power of consistency. In the fascinating film documentary *The UCLA Dynasty*, one player recalled, "There was a way to do everything. You could have taken UCLA people who played in '55, '65, '70, and '75; put them on the same team; and they would have been able to play with each other, instantly." Wooden ran his drills from the same set of 3x5 cards, with rare modifications, over the course of three decades. Drills would start and end like clockwork, the same drills performed before the national championship as at the beginning of the season so that, in the words of a star player, "By the time the games came along, they just became memorized exhibitions of brilliance."[16]

Wooden translated his "Pyramid of Success" (a philosophy of life and competition) into a detailed recipe, right down to how players should tie their shoes.[17]

Picture yourself as a star basketball player recruited to UCLA. You show up at the first practice session, ready to show your skills; to earn your spot; to run up and down the court; to slam the ball through the hoop; to leap, and jump, and spin. You sidle up next to a senior who'd earned All-American honors and wait for the coach to get the drills going. The coach comes out and opens the first moments of practice in a quiet voice, "We will begin by learning how to tie our shoes."

You look over to a couple of famous seniors, All-Americans who've already won national championships, thinking this must be some kind of freshman initiation. But no, the seniors calmly begin taking off their shoes and preparing for the shoe-tying lesson.

"First, put your socks, slowly with care, over your toes," says the coach. The seniors diligently follow instructions. "Now, move your socks up here . . . and here . . . smooth out all the wrinkles . . . nice and tight . . . take your time," the coach intones his lesson, like some sort of far-out Zen master teaching you how to make tea as a path to higher enlightenment. "Then lace your shoes from the bottom, carefully, slowly, getting each pass nice and tight . . . snug! snug! snug! snug!"

After the lesson, you ask one of the All-American seniors what that was all about, and he says, "Get a blister in a big game, and you're gonna suffer. Shoes come untied in a close game . . . well, that just never happens here." One year later, you come to practice, having helped create yet another national championship, noting the surprised looks on the freshmen's faces when the coach announces, "We will begin by learning how to tie our shoes."

Modern management dogma exhorts that an enterprise should commit frequent wholesale revolution, that it should change more on the inside than the world is changing on the outside, that it should inflict radical change upon itself, and that it should be doing so all the time. But as Lincoln said in the dark days of the American Civil War, "The dogmas of the quiet past are inadequate to the stormy present."[18] In this stormy world, we need to think anew. And that means rejecting the idea that the only path to continued prosperity lies in continuous corporate revolution. If you really want to become mediocre or get yourself killed in a turbulent environment, you want to be changing, morphing, leaping, and transforming yourself all the time and in reaction to everything that hits you. We've found in all our research studies that *the signature of mediocrity is not an unwillingness to change; the signature of mediocrity is chronic inconsistency.*

Keep in mind the premise of this study: the world is in a state of uncertainty and instability, full of rapid change and dramatic disruptions. Yet when we conducted our research *through this very lens* of extreme change and turmoil, we found that the 10X companies changed their recipes less than their comparisons. This doesn't mean 10X leaders are complacent. Productive paranoids infused with fanatic discipline and fired up by empirical creativity in pursuit of Level 5 ambitions don't have any conception of complacency. 10Xers are truly obsessed, driven people. It's just that they accomplish their huge goals by adhering with great discipline to what they know works while simultaneously worrying—for they always worry—about what might no longer work in a changing environment. When conditions truly call for a change, they respond by amending the recipe.

AMENDING THE SMaC RECIPE: PARANOID, CREATIVE CONSISTENCY

Suppose we asked you to catalogue everything in your world that's changing. How long a list would you need? Just consider a few categories:

How is the economy changing?
How are the markets changing?
How are fashions changing?
How is technology changing?
How is the political landscape changing?
How are laws and regulations changing?
How are societal norms changing?
How is your line of work changing?

The amount of change swirling about is both gigantic and, for most people, accelerating. If we tried to react to every single external change, we'd quickly find ourselves incapacitated. Most change is just noise and requires no fundamental change in ourselves.

Yet some change is *not* noise, demanding that we adjust and evolve, else we face demise, catastrophe, or missed opportunities. A great company *must* evolve its recipe, revising selected elements when conditions merit, while keeping most of its recipe intact.

In 1985, Intel faced a bleak reality in its memory-chip business (DRAMS). Japanese competitors had thrown the industry into a brutal price war, driving prices down by 80 percent in two years. Intel leadership eventually had to confront a brutal fact: the memory business no longer offered anything but bleeding and misery. Fortunately, Intel had fired bullets on another business, microprocessors, beginning in 1969, when engineer Ted Hoff put all the computing functions on a single chip. Over the subsequent 16 years, Intel had gradually built momentum in microprocessors, increasing market share, growing profits, and gaining empirical validation that microprocessors offered a huge, viable business for Intel.[19]

In a decision first made famous by Stanford Professor Robert Burgelman (the world's leading authority on Intel's strategic evolution), Andy Grove and Gordon Moore debated what to do about the declining memory-chip business. Grove *zoomed out* and posed a hypothetical question to Moore, "If we were replaced and new management came in, what would they do?"[20]

Moore thought about it for a moment, then answered, "Get out of DRAMS."

"So," said Grove, "let's go through the revolving door, come back in, shut down the memory business, and just do it ourselves."

And that's exactly what they did, throwing their full attention into the microprocessor business.

This was a very big change for Intel, yet at the same time, Intel kept intact most of the other ingredients in its recipe. Notice in the table "Intel SMaC Recipe" what did not change at the time Intel exited the memory-chip business. Certainly, if Intel had blindly stuck with memory chips, it might not have become a 10X winner. But equally true, if it had changed most of its recipe—if it had jettisoned Moore's Law, started cutting R&D, abandoned its pricing model, ruined its practice of constructive confrontation—it would not have become a 10X winner. *Both* parts of the story are important: the big exit from memory chips *and* the fact that Intel did *not* change other elements of its SMaC recipe.

Intel SMaC Recipe[21]	Changed in 1985?
1. Concentrate on integrated electronics, where all functions are supplied to the customer as irreducible units. Focus on DRAM memory chips.	Exited memory chips; shifted focus to microprocessors
2. Uphold Moore's Law, doubling the complexity of components per integrated circuit at minimum cost every 18 months to two years.	No change
3. Achieve Moore's Law by (a) increasing chip size through reducing random defects, (b) creating circuit innovations that allow for higher functional density, and (c) making circuit units smaller.	No change

4. Continuously develop the next generation of chips that create a competition-free zone. Develop chips that customers must have because Intel has a better product than the previous generation and/or has an industry standard. Maximize the benefits of the competition-free zone via a four-part cycle: (a) price high early in the cycle, (b) gain volume and drive down unit costs, (c) lower prices as competition enters and keep driving down unit costs, and (d) deploy profits into the next generation of chips to create the next competition-free zone.	No change
5. Standardize manufacturing down to the smallest details; i.e., McIntel. Think of making integrated circuits like making high-tech jelly beans.	No change
6. Maintain our reputation that "Intel Delivers." Build customers by earning their trust that we will *always* deliver on our manufacturing and price commitments. This is the secret to gaining and holding an industry standard.	No change
7. Do not attack a fortified hill; avoid markets with powerful, entrenched competitors.	No change
8. Practice constructive confrontation. Argue and debate regardless of rank, and then commit once a decision is made—disagree and commit.	No change
9. Measure everything and make visible the results.	No change
10. Do not cut R&D during recessions; use recessions to drive our technology ahead of others.	No change

The Intel case illustrates a powerful "Genius of the AND." On the one hand, a great company changes only a small fraction of its SMaC recipe at any given time, keeping the rest of it intact. On the other hand, this isn't just "incremental" change; a SMaC-recipe change is, almost by definition, a hugely significant change. By grasping this point, a 10X enterprise can achieve significant change *and* extraordinary continuity, both at the same time.

Intel's comparison case, AMD, presents a stark contrast, settling upon a recipe, then throwing it out to settle upon another recipe, then replacing it with yet another, then back again. Early in its history, AMD developed a recipe principally focused on being a second-source supplier and manufacturing chips to military specifications. Then in the early 1980s, Jerry Sanders concocted a new recipe, this time for . . . "asparagus"! Asparagus requires more up-front investment and takes longer to grow than other crops but yields higher prices. Stretching this analogy to microelectronics, Sanders and company shifted to making proprietary chips that required more up-front investment and took longer to develop yet yielded higher prices—like asparagus! AMD hung an asparagus flag outside its headquarters and took out ads proclaiming, "We're ready for the asparagus business." Then just a few years later, AMD shifted back to a second-source strategy, although it also kept some asparagus. Then it shifted to something it called the "P^3 Strategy" (platforms, process, and production). Then in yet another shift, it pursued something called "customer-centric innovation." While none of these were bad ideas per se, in switching from one recipe to another, inflicting frequent wholesale change upon itself, AMD never gained long-term momentum.[22]

So, how does a 10X company know when it's time to amend its recipe, presuming it has a really good one? With a concrete recipe in hand, it can explicitly consider the recipe's ingredients in the context of changes in the environment. It can examine the empirical evidence. What are the brutal facts? Not opinions, but *facts*. What bullets have we fired? What have they hit? The Intel case illustrates how firing bullets can give you a hedge against an uncertain future, so that you might have a ready-made amendment ready to go when the world changes. Intel didn't react to the memory-business disruption by inventing the microprocessor; it had been firing bullets for more than a decade, proving itself in microprocessors.[23]

There are two healthy approaches to amending the SMaC recipe: (1) exercising empirical creativity, which is more internally driven, and (2) exercising productive paranoia, which is more externally focused. The first involves firing bullets to discover and test a new practice before making it part of the recipe. The second employs the discipline to *zoom out* to perceive and assess a change in conditions, then to *zoom in* to implement amendments as needed.

10Xers employ both approaches, although the emphasis can vary depending upon the situation. In the Intel case, empirical creativity came first (firing bullets on the microprocessor) and then productive paranoia kicked in when the memory-chip business became untenable. Microsoft's move to embrace the Internet in the 1990s illustrates how productive paranoia might provide the initial spark for an amendment.

Prior to 1994, Microsoft built its recipe around the stand-alone personal computer as the center of the universe. Then in January 1994, a 25-year-old Microsoft engineer named James J. Allard sounded an alarm, pointing out that two new systems were being added to the Internet with every passing minute, with a new network connecting up every 40 minutes. A month later, one of Microsoft's technical generals visited Cornell University, and seeing firsthand how all the kids were connected to the Internet, followed up with an email to Bill Gates, "Cornell is WIRED!" Sensing a change in conditions, just like David Breashears on Everest, Gates *zoomed out*. In fact, Gates had a *zoom-out* mechanism already in place; he'd set aside an entire week each year to go away for intense reading and reflection, his "Think Week." Gates dedicated his April 1994 Think Week to the Internet. He also stimulated his team to *zoom out*, calling a retreat of the Microsoft brain trust to assess the threat. What are the facts? Does this require a major change? Is this real or is it hype? Are we threatened? The discussions, debates, and yelling matches persisted over the course of months. Finally, Microsoft came to see that the Internet did indeed represent a fundamental change in the environment and a serious threat; Microsoft would need to fully embrace a wired world.[24]

Then Microsoft *zoomed in*. Gates wrote a memo, eight pages, single-spaced, entitled "The Internet Tidal Wave," in which he described his own evolution, having "gone through several stages of increasing my views as to [the Internet's] importance." He then redirected Microsoft to the Internet, pushing his teams to "go overboard on Internet features" and sending more than five hundred programmers on a speed march to develop a browser that would later be known as Internet Explorer.[25] The memo became the stuff of legend, the story of how a visionary founder revolutionized his company, turning the battleship 180 degrees overnight; and it makes for fascinating reading.

Yet just as Intel handled its transition to microprocessors, Microsoft kept intact most of the recipe that had made it successful prior to the rise of the Internet. Microsoft did not abandon its focus on software. It did not abandon its belief in standards. It did not abandon its approach of launching imperfect products and then improving them. It did not abandon its price-for-volume strategy. It did not abandon its commitment to open systems. It did not abandon the practice of internal yelling matches, the testing ground for letting the best ideas win. It did not abandon Windows. It did not abandon applications. It was a huge change for Microsoft to embrace the Internet, and yet most of Microsoft's recipe remained intact. Did Microsoft make a big change to its recipe? Yes. Did Microsoft keep most of its recipe intact? Yes. Again, 10Xers reject the choice between consistency and change; they embrace consistency and change, both at the same time.

CONSISTENCY AND CHANGE: THE GREAT HUMAN TENSION

When the framers of the United States Constitution convened in Philadelphia in 1787, they wrestled with a profound question, how to create a practical framework that would be both flexible and durable. Go too far in one direction, putting in too many specific strictures, and the Constitution would become either a straitjacket or irrelevant. The framers couldn't possibly predict how the world would change, having no capacity even to anticipate or envision automobiles, airplanes, talk radio, cable news, the Internet, the Civil Rights Movement, nuclear weapons, birth-control pills, the rise of the Soviet Union, the fall of the Soviet Union, jazz, multimillion-dollar athletes on strike, American dependence on foreign oil, or 9/11. Go too far in the other direction, providing only broad and general guidelines, and the Constitution would lack "teeth" and fail to provide the practical guideposts that meld a diverse group of people and individual states into a single union. There must be a coherent, consistent, enduring framework holding the enterprise together, preventing disintegration into a squabbling group of independent little countries.

So, they came up with an ingenious invention, the amendment mechanism. One of the first of its type in human history, the mechanism would allow the Constitution to evolve organically, enabling future generations to make adjustments when situations arose that the founders could not possibly envision. Equally important, they designed the mechanism to ensure stability, creating a very high hurdle for change. After the first 10 Amendments (the Bill of Rights) in 1791, there were *only 17* amendments in the next 220 years. The framers had made amendments rare by design, requiring a two-thirds majority in the House of Representatives, a two-thirds majority in the Senate, and then ratification by three-fourths of the individual states. Think of everything that happened between 1791 and 2011, and yet the Constitution was amended only 17 times. The authors of the Constitution clearly understood that change must be possible, but they also understood that a great nation must have a consistent framework to work from *especially* in a radically changing and utterly unpredictable world.[26]

Any enterprise, whether a company, society, nation, church, social venture, school, hospital, military unit, orchestra, team, or any other human organization, faces a constant struggle to find the balance between continuity and change. No human enterprise can succeed at the highest levels without consistency; if you bring no coherent unifying concept and disciplined methodology to your endeavors, you'll be whipsawed by changes in your environment and cede your fate to forces outside your control. Equally true, however, no human enterprise can succeed at the highest levels without productive evolution.

We came to see the way 10X enterprises reconciled this great human tension as similar to the way the framers thought about the Constitution and the amendment mechanism. You need concrete rules of the road to guide decisions, providing a coherent framework and consistency over time. And you need to take the time to get those rules right, building them upon a savvy understanding about what actually works. In 1787, the new nation sent some of its very best people to Philadelphia for four months to work out the details of the Constitution. The Declaration of Independence provided the idealism ("We hold these truths to be self-evident"), but the Constitution needed to realistically take into account how people and power actually work, about the undying forces of self-interest, about the necessity of checks and balances, about the dangers of reactionary masses, about the value of compromise. And it needed a mechanism for change.[27]

> Changes to a solid and proven SMaC recipe are like amendments to the Constitution: if you get the recipe right, based on practical insight and empirical validation, it should serve you well for a very long time; equally important, fundamental changes must be possible. Continually question and challenge your recipe, but change it rarely.

Greatness comes to those who keep moving forward, figuring out what works, driving down Moore's Law, advancing the Southwest Airlines model across the country, cracking the code for EPO, marching relentlessly to make Windows a standard, making computers and MP3 players that we'd want for ourselves. Those who spend most of their energy "reacting to change" will do exactly that, expend most of their energy reacting to change. In a great twist of irony, those who bring about the most significant change in the world, those who have the largest impact on the economy and society, are themselves enormously consistent in their approach. They aren't dogmatic or rigid; they're disciplined, they're creative, they're paranoid. They're SMaC!

SMaC

KEY POINTS

▶ SMaC stands for Specific, Methodical, and Consistent. The more uncertain, fast-changing, and unforgiving your environment, the more SMaC you need to be.

▶ A SMaC recipe is a set of durable operating practices that create a replicable and consistent success formula; it is clear and concrete, enabling the entire enterprise to unify and organize its efforts, giving clear guidance regarding what to do and what not to do. A SMaC recipe reflects empirical validation and insight about what actually works and why. Howard Putnam's 10 points at Southwest Airlines perfectly illustrates the idea.

▶ Developing a SMaC recipe, adhering to it, and amending it (rarely) when conditions merit correlate with 10X success. This requires the three 10Xer behaviors: empirical creativity (for developing and evolving it), fanatic discipline (for sticking to it), and productive paranoia (for sensing necessary changes).

▶ All but one of the comparison cases also had solid recipes during their best years, yet they lacked the discipline to implement them with creative consistency, often making reactionary lurches in response to turbulent times.

▶ Amendments to a SMaC recipe can be made to one element or ingredient while leaving the rest of the recipe intact. Like making amendments to an enduring constitution, this approach allows you to facilitate dramatic change and maintain extraordinary consistency. Managing the tension between consistency and change is one of the great challenges for any human enterprise.

▶ There are two healthy approaches to amending the SMaC recipe: (1) exercising empirical creativity, which is more internally driven (fire bullets, then cannonballs), and (2) exercising productive paranoia (*zoom out*, then *zoom in*), which is more externally focused.

UNEXPECTED FINDINGS

▶ It is possible to develop specific, concrete practices that can endure for decades—SMaC practices.

▶ Once they had their SMaC recipes, the 10X cases changed them only by an average of 15 percent (compared to 60 percent for the comparison cases) over their respective eras of analysis, and any given element of a 10X recipe lasted on average for more than two decades. This is a stunning finding, given that all the companies in the study, 10X cases and comparisons alike, faced rapid change and unrelenting uncertainty.

▶ Far more difficult than implementing change is figuring out what works, understanding why it works, grasping when to change, and knowing when not to.

ONE KEY QUESTION

▶ What is your SMaC recipe and does it need amending?

7

RETURN ON LUCK

> "Look, if you had one shot, or one opportunity
> To seize everything you ever wanted in one moment
> Would you capture it? Or just let it slip?"
>
> —Marshall Bruce Mathers III, "Lose Yourself"[1]

In May 1999, Malcolm Daly and Jim Donini stood three thousand feet up an unclimbed face on Thunder Mountain in Alaska, only a few hundred feet below the summit. Daly offered to let Donini go first on the rope to experience the joy of reaching the summit first, but Donini said, "No, you keep it, you are the one who deserves the gift."[2]

Less than an hour later, Daly would be dangling at the end of the rope, legs shattered, just beginning an epic fight for his life, a life that would be forever transformed by losing one of his feet.

Daly climbed toward the summit, swinging his ice axe like a giant claw, kicking knife-like spikes attached to his boots (called crampons) into the ice, moving methodically up the near-vertical wall. He dragged the safety rope (knotted to his waist harness) along behind him, while Donini remained anchored to the wall, feeding the rope through a friction device that would snap tight if the rope suddenly jerked, like a car seatbelt that would seize tight in a crash. The plan: Daly would climb to the summit ridge, placing protection points along the way (mainly "ice screws" twisted into frozen solid sheets of ice); anchor himself to the top of the mountain; and then hold the safety rope while Donini climbed up to meet him.

With only about 15 feet of steep climbing to go, Daly reached a section of rock where he could place no protection. No problem, though, the final few feet of climbing looked easy. Daly placed his left hand on a big jut of rock, groping about with his right hand, looking for another hold, thinking to himself, "Gosh, this next move is it and there are no more moves on the route. We are essentially up."

Something gave way.

He fell.

Ten feet.

Twenty feet.

Ice screws ripped out.

Forty feet.

A hundred feet.

Still falling!

The rope whipped, the gear clangled as Daly bounced and flew.

He smashed into his partner, puncturing Donini's right thigh with the pointed teeth of his crampons.

Daly hurtled past.

Still falling.

Sixty more feet.

Something sharp sliced the rope. Ten of twelve core strands of rope severed right through. If the remaining two were to break . . .

Daly cratered into the mountainside. The two remaining strands of cord, less than two millimeters thick, stretched but didn't break. Daly stopped, a crumpled lump.

"Malcolm, Malcolm, are you okay? Are you alive?" yelled Donini, thinking that Daly must be dead.

Daly didn't respond.

Donini kept yelling. No response.

Then finally, Daly regained consciousness. Blood dripped from his scalp. He looked at his lower legs and feet, shattered with compound fractures; feet flopping around, useless. Daly felt the ends of busted bones rubbing together.

Donini descended to Daly, and they tried to engineer a self-rescue but soon realized that any movement could worsen the compound fractures and Daly might bleed to death. Daly told Donini, "You have to go get a rescue." After anchoring Daly to the wall, Donini took off on a three-thousand-foot solo descent.

Within minutes after Donini reached base camp at the bottom of the mountain, he heard something quite unexpected: his friend Paul Roderick of the Talkeetna Air Taxi (an expedition-support service) just happened to be flying by that particular valley at that exact moment. Donini waved him down, and Roderick flew Donini directly to the ranger station; a plan to rescue Daly began immediately, many hours sooner than if Donini had needed to hike out to the station. Those hours proved pivotal. By the time the rescue was organized, impending storms threatened to curtail the attempt. Racing the weather, a helicopter flew up to Daly's perch, and a rescue pilot hanging from a cable below the chopper swung into the mountainside and plucked Daly off the mountain.

Four hours later, a huge storm enveloped the mountain and raged on for 12 days.

LUCK OR SKILL?

Now, ask yourself, what role did luck play in this story? There's the bad luck of Daly's seemingly solid stance inexplicably giving way, sending him hurtling into the abyss. And there is lots of good luck. The sliced rope wasn't cut all the way through. Daly didn't die in the fall. He didn't kill Donini on the way down. Donini reached base camp just as the airplane happened to fly by. And had everything taken just five hours longer, Daly would not have survived.

But now, let's add some other pieces to the story.

Malcolm Daly had prepared well in advance. He drew on tremendous physical reserves and wilderness experience, layers of fitness and strength built by thousands of hours of rigorous training—biking, climbing, running, skiing, and mountaineering. He'd also prepared mentally, reading survival literature "just in case" he ever ended up in a desperate battle for his life. In fact, just days before the climb, he'd been reading about Ernest Shackleton and his mission to rescue himself and his men from Elephant Island, Antarctica, in 1916. Daly learned from his preparation that wallowing in your misfortune increases risk. "I loved my feet," he later reflected. "[But] there was nothing I could do that would affect the outcome of my feet, other than worry about them too much and add that level of stress and then perhaps I could hurt my chances of survival. So I put that thought on a shelf."

Daly made a plan to live, what he later described as a *decision* to live. He had to stay warm, not go hypothermic. So, he set forth a regimen: do 100 windmills on one arm, swinging it around in full 360-degree circles; then 100 on the other arm; then 100 stomach crunches; then repeat without stopping, keeping his mind focused, counting precisely, not "approximately" 100, but *exactly* 100. He tired but kept a regimen, dropping the sets to 50, then eventually 20, but always with the regimen. That Daly had the stamina and tenacity to keep this up for *44 hours* is certainly not luck.

He had the right partner in Jim Donini, as Daly had always chosen his partners with great care, knowing that the ultimate hedge against danger and uncertainty is whom you have on the mountain with you. Donini had logged thousands of days in the mountains, from Patagonia to the Himalaya, capturing some of the most coveted first ascents in climbing history, and he was one of only a handful of people in the world with the skill to descend three thousand feet *solo*, without a single misstep, despite having a punctured thigh.[3]

When the rescue began, Daly prepared for the helicopter by cutting open the pack into which he'd stuffed his broken feet so they'd pull out with ease; slicing away his bloody, frozen leg coverings; and chipping away any residual ice that might have frozen something to the wall. He knew to take these steps because he'd studied helicopter rescues. And he was ready.

Which brings us to perhaps the most significant element in Daly's survival: he'd developed relationships with people who loved him and who would risk their lives for him. The rescue leader who swung in from the helicopter, Billy Shot, was a longtime friend. When Shot swung onto the snow slope, his radio communications went awry, which would have normally meant an automatic abort. But Shot knew he *had* to get his friend—his friend!—off the mountain before the storm, so he made an on-the-spot switch to hand signals. Clawing at the snow with ice tools, he gouged and scampered up to Daly, snapped him onto the cable, then signaled the helicopter to whoosh them away from the mountain. Shot held Daly in a huge bear hug. As they dangled from a cable thousands of feet in the air, Shot sported a huge smile. "You know who I am?" Daly shook his head, unable to see his rescuer's face. Then Shot lifted his faceplate. "It's Billy Shot!" Daly's friend had come to save his life and deliver him to safety. Luck clearly played a role in Daly's survival, but luck didn't save Daly in the end. People did.

WHAT ROLE LUCK?

The very nature of this study—thriving in uncertainty, leading in chaos, dealing with a world full of big disruptive forces that we cannot predict or control—led us to the fascinating question, "Just what is the role of luck?" And how, *if at all*, should luck factor into developing our strategies for survival and success? Perhaps everything we've studied and written about, what leaders and people *do*, accounts for the difference between only 1X and 2X success, and luck accounts for the difference between 2X and 10X. Perhaps the 10X winners really were just much luckier than the comparisons. Or . . . perhaps not.

We decided to undertake an analysis of luck, asking three basic questions:

1. Is luck a common or rare element in the histories of the 10X and comparison cases?

2. What role, if any, does luck play in explaining the divergent trajectories of 10X and comparison cases?
3. What can leaders *do* about luck to help them build great companies on a 10X journey?

But first, we had to develop a rigorous and internally consistent method for analyzing the topic, beginning with a clear definition of a luck event. We realized that people think about luck in imprecise ways, captured in common phrases like "Luck is where preparation meets opportunity" or "Luck is the residue of design" or even "The harder I work, the luckier I get." None of these oft-repeated phrases are precise enough to actually analyze the role of luck, so we constructed a definition that would allow us to engage directly with the topic, focusing on identifying specific luck events.

We defined a luck event as one that meets three tests: (1) some significant aspect of the event occurs largely or entirely independent of the actions of the key actors in the enterprise, (2) the event has a potentially significant consequence (good or bad), and (3) the event has some element of unpredictability.

All three parts of the definition are important. Some significant aspect of the event must happen *largely or entirely independent of the actions of the key actors.* For example, Daly and Donini didn't cause Paul Roderick to fly by in his airplane at just the right moment; it was a huge luck event, especially given the time pressure to get Daly off the mountain before a looming storm. The event must have a potentially *significant consequence* (good or bad); consider the two uncut rope strands that stopped Daly's fall. The event must have some element of *unpredictability*; Daly didn't foresee that his seemingly solid stance would give way, sending him hurtling on a two-hundred-foot fall.

Yet notice how other details of the Daly/Donini story do *not* qualify as luck. Daly's 44-hour marathon of sit-ups and arm circles was an act of sheer will and incredible fitness. Donini's successful three-thousand-foot solo descent down the face of the mountain was a matter of skill and experience. Daly's friends would risk their lives to save his, not because of luck, but because they knew he would've done the same for them.

Our definition of luck leaves unaddressed the possible explanations for the event's ultimate cause. Whether the luck event results from randomness, accident, complexity, Providence, or any other force, doesn't matter for the sake of our analysis. You could look at the two unbroken strands of Daly's rope as pure chance, or as a miracle. As long as an event meets the three dimensions of our definition, whatever its cause, it qualifies as a luck event.

We developed a method that considered the significance of each luck event to account for the fact that some events have greater impact than others, taking care to be consistent in our analysis within each pair. In the table "Luck-Coding Example: Amgen versus Genentech," we've listed 7 representative luck events for each company (from the combined total of 46 luck events we identified for this pair) to illustrate.

> Analyzing luck is difficult, and perhaps novel. By applying a *consistent methodology to both members of each matched pair,* we were able to use evidence-based analysis to attack this elusive topic, focusing on the question, "Did the 10X company get more good luck, or less bad luck, than the comparison company?"

LUCK-CODING EXAMPLE

AMGEN VERSUS GENENTECH

The 14 luck events described here are a representative list of the 46 luck events analyzed for this matched pair.

Amgen

Luck Event	Assessment
1981: A Taiwanese scientist named Fu-Kuen Lin just happened to see (and respond to) a small help-wanted classified advertisement placed by Amgen.[4] Amgen could neither control who saw the ad nor predict that one of the respondents would be a genius with the ferocity to persist against all odds and skeptics to lead the EPO-gene breakthrough. Amgen's decision to take out a classified ad isn't luck; that Fu-Kuen Lin happened to see the ad at the precise moment he was looking for a job opportunity is luck.[5]	Good luck High importance
1982: The biotechnology industry experienced a downturn, which impacted investor sentiment and funding options for the fledgling company; this was potentially significant for Amgen given that it planned to go public soon thereafter.[6]	Bad luck Medium importance
1983–89: Amgen isolated the gene for EPO, which it likened to "finding a sugar cube in a lake one mile wide, one mile long and one mile deep." EPO proceeded through clinical trials and FDA approval. Creating a successful new product in biotechnology always involves some element of luck, no matter how skilled the R&D people; there's always a chance it will not make it from concept all the way through clinical trials to FDA approval.[7]	Good luck High importance
1987: A rival company, Genetics Institute, was issued a patent that circumvented Amgen's proprietary technology for producing EPO. Amgen had cracked the code on how to make bioengineered EPO; Genetics Institute had gained a patent on so-called "natural" EPO made from human urine. An article in *Nature* summed up, "Genetics Institute has a claim on the final destination and Amgen on the only way of getting there."[8] This unexpected event imperiled Amgen's ability to fully capitalize on its breakthrough.	Bad luck High importance
1991: The Court of Appeals for the Federal Circuit reversed court decisions that Amgen had previously lost in its dispute with Genetics Institute, and the U.S. Supreme Court declined to hear a further appeal by Genetics Institute, giving Amgen complete victory. That Amgen mounted a smart and ferocious legal defense isn't luck; the luck here stems from the fact that how a court rules and whether the Supreme Court agrees to hear an appeal cannot be determined by the company. The outcome was a big surprise to many observers who believed that Amgen could not win its case and should seek a settlement.[9]	Good luck High importance

1995: The anti-obesity gene, leptin, didn't make it through all the probability gates to a successful product. The market potential was gigantic; if the product were to work, people would be able to take a pill to cut their appetite and reduce their weight. Amgen's initiative didn't affect patients sufficiently to merit continued development, and Amgen discontinued clinical trials.[10]	Bad luck Medium importance
1998: MGDF (Megakaryocyte Growth and Development Factor), which reduced platelet loss during chemotherapy and was regarded as a likely "home run" product, didn't make it through all the probability gates. It could have been a quarter-of-a-billion-dollar product by 2000, but clinical trials showed that some patients developed antibodies that neutralized its effectiveness.[11]	Bad luck Medium importance

Genentech

Luck Event	Assessment
1975: Financier Robert Swanson and molecular biologist Herbert Boyer happened to be in the right place (the San Francisco Bay Area) at the right time (just as scientific advancements made gene splicing viable) when they met for the first time. They hit it off, becoming fast friends, and realized that a confluence of forces (the rise of venture capital and the advancement of gene-splicing technology) made possible the creation of the first biotechnology company in history.[12] That they started a company isn't luck; that they happened to be in precisely the right place at precisely the right time to be first is luck.	Good luck High importance
1980: *Time* magazine dedicated an entire page to Genentech's impending public stock offering.[13] The offering far exceeded expectations, becoming one of the first supernova public offerings in modern business history—a Netscape or Google IPO of its time—with its stock price rising more than 150% ($35 to $89) in less than a day.[14] That Genentech had a successful IPO isn't luck; that its stock jumped 150% in a day was unexpected, uncontrollable, and significant luck.	Good luck Medium importance
1982: Genentech became the first company to succeed at applying gene splicing to create a commercially viable recombinant-DNA drug (human insulin) that was approved by the FDA.[15] That Genentech figured out how to splice genes isn't luck; that no one else got there first is luck. That Genentech developed a product isn't luck; that it made it through clinical trials to FDA approval is luck.	Good luck High importance

(*continued on next page*)

Luck Event	Assessment
1982: The biotechnology industry experienced a downturn, impacting investor sentiment; shares fell to less than $35 from the IPO high of $89, raising the cost of capital. Markets are always uncontrollable and unpredictable. The downturn was potentially significant as Genentech had less than $1 million in profits and depended upon access to equity capital to fund breakthrough R&D.[16]	Bad luck Medium importance
1987: Genentech's t-PA discovery, trade-named Activase, made it through clinical trials to win FDA approval. The potential market was huge, as it could be used to stop heart attacks in the early stages.[17] The chairman of medicine at Harvard Medical School said that t-PA would "do for heart attacks what penicillin did for the treatment of infections." [18] Seen as biotech's first blockbuster, Activase was proclaimed as the "most successful new drug ever launched" and seen as a candidate for turning Genentech into the first billion-dollar-revenue biotechnology company, "biotech's first superstar." [19]	Good luck High importance
1989: The *New England Journal of Medicine* published an article that challenged the effectiveness of t-PA relative to more conservative strategies and alternative treatments.[20] Other studies also challenged t-PA.[21] Genentech could not control studies done outside its own walls; the prestige of the *New England Journal of Medicine* increased the significance of the event.	Bad luck High importance
1993: A study called GUSTO found that Genentech's t-PA, contrary to earlier studies, did save more lives than alternative treatments; t-PA regained market support and jumped to having 70% market share.[22] That Genentech sponsored the GUSTO study isn't luck; that the study validated t-PA had an element of luck, as there's always the possibility that such a study will not produce hoped-for results.	Good luck High importance

We became increasingly excited by this analysis, curious to see just what the data would show. After all, to our knowledge, no one had ever taken on the topic of luck in this way, and we didn't know what the evidence would yield. Using our definition, we identified and systematically coded 230 significant luck events for the 10X and comparison cases. All the companies had good luck and bad luck—luck happens, a lot—but does luck play a *differentiating* role, an explanatory role, a definitive role in creating 10X success?

To get at this question, we looked through multiple lenses. (See *Research Foundations: Luck Analysis.*) First, we considered whether the 10X cases got substantially more *good* luck than the comparison cases. As a general rule, the answer was *no.* The 10X cases averaged seven significant good-luck events and the comparison cases averaged eight significant good-luck events across the era of analysis, with no evidence that the 10X cases got substantially more good-luck events than the comparisons.

Matched Pairs	Number of Significant Good-Luck Events	
	10X Case	*Comparison Case*
Amgen and Genentech	10	18
Biomet and Kirschner	4	4
Intel and AMD	7	8
Microsoft and Apple	15	14
Progressive and Safeco	3	1
Southwest and PSA	8	6
Stryker and USSC	2	5
Average	7	8
Total	49	56

Then we considered: did the comparison cases get more *bad* luck than the 10X cases? As a general rule, the answer was no; the analysis showed similar levels of bad luck, each group averaging about nine bad-luck events.

Matched Pairs	Number of Significant Bad-Luck Events	
	10X Case	*Comparison Case*
Amgen and Genentech	9	9
Biomet and Kirschner	7	4
Intel and AMD	14	11
Microsoft and Apple	9	7
Progressive and Safeco	8	10
Southwest and PSA	13	13
Stryker and USSC	5	6
Average	9.3	8.6
Total	65	60

Then we considered whether a single luck event—a big "luck spike"—could be so huge as to explain nearly all the success of a 10X company relative to its comparison. But in only one pair, Intel versus AMD, did we see a huge luck spike on one side of the pair (IBM's selecting the Intel microprocessor for its personal computer) *without* a corresponding comparable luck spike on the other side of the pair. Even in this case, Intel's three *decades* of sustained success cannot be fully explained by this individual luck event, especially given the company's earned reputation that "Intel Delivers," solidly in place since the early 1970s.[23] As a general finding, both the 10X cases and the comparisons each got some big good-luck events and some big bad-luck events; the evidence does not support the hypothesis that the 10X cases won because of one gigantic piece of luck that dwarfed everything else.

Finally, we analyzed the time distribution of luck, wondering if perhaps the 10X cases got their good luck early, while the comparison cases got their bad luck early, before they had a chance to fully establish themselves. Perhaps getting an outsized share of good luck early set the 10X cases on a permanently superior path. But again, we found no significant difference. The 10X cases did not systematically have more good luck early, and the comparisons did not have more bad luck early. 10Xers won not because they had better early advantages or superior early luck. As a general rule, they had neither.

Throughout our analysis, we were very careful to distinguish between luck and outcomes. An enterprise can get bad luck yet create a good outcome, and equally, a company can squander good luck and get a bad outcome. The real difference between the 10X and comparison cases wasn't luck per se but what they *did* with the luck they got.

Adding up all the evidence, we found that the 10X cases were *not* generally luckier than the comparison cases. The 10X cases and the comparisons *both* got luck, good *and* bad, in comparable amounts. The evidence leads us to conclude that luck does not cause 10X success. People do. The critical question is not "Are you lucky?" but "Do you get a high *return on luck*?"

WHO IS YOUR BEST LUCK?

In 1998, Amgen chairman Gordon Binder gave a speech at the Newcomen Society in which he identified "perhaps the defining moment in the Amgen story." And what did he pick? Early venture funding? Amgen's IPO? FDA approval of blockbuster EPO? Some other major product? Nope. The "defining moment" came when Taiwanese scientist Fu-Kuen Lin just happened to see (and respond to) a help-wanted ad.[24]

When George Rathmann drove into the company parking lot one morning before dawn in 1982, he spotted lights glowing in a lab building. "Someone must have left them on last night by mistake," he thought to himself. When he entered the lab to turn out the lights, he found Lin toiling away; he'd been there all night. Unassuming, ferociously patient, and relentless in his work, Lin attacked the problem of cloning the EPO gene, logging 16-hour days for nearly two years non-stop. The problem proved so difficult that people avoided Lin's seemingly quixotic quest. "My assistant was told by the other associates, 'What a dummy you are to work with this guy on a project that is going nowhere,' " reflected Lin. What if Lin had never seen the ad? What if he'd taken a job elsewhere? Would Amgen have created the first billion-dollar biotechnology blockbuster?[25]

We tend to think of luck as a "what" variable—the plane flies by at the right moment, your IPO becomes much more successful than expected, etc. But one of the most significant forms of luck comes not as "what" but in the form of *who*. In a family business, for example, there's a significant amount of luck in whether a son or daughter has the right stuff to lead a company to greatness; Progressive began as a small family business in Cleveland, Ohio, and the family owners got a remarkable 10Xer son in Peter Lewis, who took over the company in 1965.[26]

This research project began with the premise that we live in an environment of chaos and uncertainty. But the environment doesn't determine why some companies thrive in chaos and why others don't. *People* do. People are disciplined fanatics. People are empirical. People are creative. People are productively paranoid. People lead. People build teams. People build organizations. People build cultures. People exemplify values, pursue purpose, and achieve big hairy audacious goals. Of all the luck we can get, people luck—the luck of finding the right mentor, partner, teammate, leader, friend—is one of the most important.

HIGH ROL: RETURN ON LUCK

Why did Bill Gates become a 10Xer, building a truly great software company in the personal computer revolution? Through one lens, you might see Bill Gates as incredibly lucky. He just happened to have been born into an upper-middle-class American family that had the resources to send him to a private school. His family enrolled him at Lakeside School in Seattle, which had obtained a teletype connection to a computer upon which he could learn to program, something unusual for schools in the late 1960s and early 1970s. He just happened to have been born at the right time, coming of age just as the advancement of microelectronics made the personal computer inevitable; born 10 years later, or even 5 years later, he would have missed the moment. His friend Paul Allen just happened to see a cover story in the January 1975 issue of *Popular Electronics* titled "World's First Microcomputer Kit to Rival Commercial Models." It was about the Altair, designed by a small company in Albuquerque. Gates and Allen had the idea to convert the programming language BASIC into a product that could be used on the Altair, which would put them in position to be the first to sell such a product for a personal computer. Gates went to college at Harvard, which just happened to have a PDP-10 computer upon which he could develop and test his ideas.[27] Wow, Gates was really lucky, right?

Yes, Gates was lucky, but luck is not *why* Gates became a 10Xer.

Consider the following questions:

Was Gates the *only* person of his era who grew up in an upper-middle-class American family?

Was Gates the *only* person born in the mid-1950s who attended a secondary school with access to computing?

Was Gates the *only* person who went to a college with computer resources in the mid-1970s?

Was Gates the *only* person who read the *Popular Electronics* article?

Was Gates the *only* person who knew how to program in BASIC?

No, no, no, no, and no.

Lakeside might have been one of the first schools to have a computer that students could access during those years, but it wasn't the only such school.[28] Gates might've been a math and computer whiz kid at a top college that had computers in 1975, but he wasn't the only math and computer whiz kid at Harvard, Stanford, Princeton, Yale, MIT, Caltech, Carnegie Mellon, Berkeley, UCLA, Chicago, Georgia Tech, Cornell, Dartmouth, USC, Columbia, Northwestern, Penn, Michigan, or any number of other top colleges with comparable or even better computer resources. Gates wasn't the only person who knew how to program in BASIC; the language had been developed by professors at Dartmouth a decade earlier, and it was widely known by 1975, used in academics and industry.[29] And what about all the master's and PhD students in electrical engineering and computer science who had even *more* computer expertise than Gates on the day the *Popular Electronics* article appeared? Any of them could have decided to abandon their studies and launch a personal computer–software company, as could have computer experts already working in industry and academia.

But how many of them disrupted their life plans (and cut their sleep to near-zero, inhaling food as fast as possible so as not to let eating interfere with work) to throw themselves into writing BASIC for the Altair? How many of them defied their parents, dropped out of college, and moved to Albuquerque—Albuquerque! New Mexico!—to work with the Altair? How many of them got BASIC for the Altair written, debugged, and ready to ship before anyone else?[30] Thousands of people *could* have done the exact same thing as Gates, at the exact same time, *but they didn't.*

> The difference between Bill Gates and similarly advantaged people is not luck. Yes, Gates was lucky to be born at the right time, but many others had this luck. And yes, Gates was lucky to have the chance to learn programming by 1975, but many others had this same luck. Gates *did* more with his luck, taking a confluence of lucky circumstances and creating a huge *return* on his luck. And this is the important difference.

When we first started working on the luck analysis, a number of our colleagues and associates said, "If you can't cause luck—if luck is something, by definition, that's out of your control—why spend time thinking about it and studying it?" True, luck happens, good and bad, to everyone, whether we like it or not. But when we look at the 10Xers, we see people like Gates who recognize luck and seize it, leaders who grab luck events and make more of them than others do. It's the 10X ability to get a high return on luck at pivotal moments that distinguishes them and this has a huge multiplicative effect. They *zoom out* to recognize when a luck event has happened and to consider whether they should let it disrupt their plans. Imagine if Bill Gates had said to Paul Allen after seeing the *Popular Electronics* article, "Well, Paul, I'm kinda focused on my studies here at Harvard right now. Let's wait a few years and then I'll be ready to start."

Look at the diagram entitled "Don't Confuse Luck with Return on Luck," which we'll use as an organizing framework for the middle of

this chapter. Everyone gets luck, good and bad, but 10X winners make more of the luck they get. The Bill Gates story illustrates the upper-right quadrant, getting a great return on good luck.

We've encountered two extreme views on the topic of luck. The first sees luck as the dominant explanation for abnormal success, holding that big winners are merely the fortunate beneficiaries of a series of lucky coin flips; after all, if you put a million monkeys in a room flipping coins, some monkey will eventually garner a string of 50 heads in a row just by random chance. In this view, people like Bill Gates are the lucky people who just happened to flip 50 heads in a row. The second extreme view claims that luck plays no role, our success and survival deriving entirely from skill, preparation, hard work, and tenacity. Those who espouse this view dismiss the undeniable fact of luck: "Luck played no role in my success; I'm just really good." In this view, Bill Gates could have become *Bill Gates* even if he'd grown up as a peasant in Communist China during the Cultural Revolution.

Our research doesn't support either extreme. On the one hand, we cannot deny the fact of luck or deny that some people start from a more fortunate place in life. On the other hand, luck by itself does not explain why some people build great companies and others don't. Our unit of analysis isn't a single event or a short moment in time; we examine great companies that sustained excellent performance for a minimum of 15 years and the leaders who built them. Across all the research we've conducted for this book and our previous books regarding what makes companies great (which has involved investigating the histories of 75 major corporations), we've never found a single instance of sustained performance due simply to pure luck. Yet also true, we've never studied a single great company devoid of luck events along its journey. Neither extreme—it's *all* luck or luck plays *no* role—has the evidence on its side. A far better fit with the data is a synthesizing concept, return on luck.

Getting a high return on luck requires throwing yourself at the luck event with ferocious intensity, disrupting your life, and not letting up. Bill Gates didn't just get a lucky break and cash in his chips. He kept pushing, driving, working—staying on a 20 Mile March; firing bullets, then big calibrated cannonballs; exercising productive paranoia to avoid the Death Line; developing and amending a SMaC recipe; hiring great people; building a culture of discipline; never deviating from his monomaniacal focus—and sustained his efforts for more than two decades. That's not luck; that's return on luck.

SQUANDERING LUCK: POOR RETURN ON GOOD LUCK

When we turn to the comparison companies, we see a substantial number of good luck events but a generally poor overall return on luck. Some of the comparison cases got extraordinary sequences of good luck yet showed a spectacular ability to fritter it away.

In the mid-1990s, perennial also-ran AMD experienced a series of good-luck events. First, a federal jury cleared the company to essentially clone Intel microprocessors, a huge court victory that gave AMD a chance to take advantage of a rising customer tide against Intel's power. Computer makers desperately wanted an alternative source for microprocessor chips, chafing at being beholden to powerful Intel. AMD developed its K5 chip, going head-to-head with Intel's Pentium chip, and customers began to make commitments to AMD. Then with AMD building momentum, clocking sales records, and lessening Intel's power, came a huge stroke of good luck: IBM halted shipments of computers that used Intel's Pentium chip due to a highly publicized glitch that caused a rounding error in certain rare calculations. Intel eventually announced a $475 million charge against earnings to replace Pentium chips for its customers. And all this happened just as the technology boom fueled huge growth in chip demand.[31]

And what did AMD do with all this good luck?

"AMD developed a rip in its mainsail, and we didn't catch the wind," wrote Sanders in his 1995 annual report. "The rip in our mainsail was our tardiness in bringing to market our fifth-generation AMD-K5 microprocessor." The K5 project slipped months behind schedule, and customers began to turn back to Intel, driving AMD's microprocessor sales down 60 percent. By the time AMD solved its K5 problems, Intel had moved on to the next generation of microprocessors.[32] AMD appeared to be out of the race, again.

Then, against all reasonable odds, AMD got another *two* lucky breaks. First, a small company named NexGen had developed a working clone of Intel's next-generation microprocessor, and—lucky for AMD—NexGen had run short of cash, forcing it to seek a friendly buyer. AMD purchased NexGen and in one step put itself back in the game. In fact, the resulting AMD-K6 appeared to be faster and cheaper than Intel's Pentium Pro when running Windows. Second, the entire industry took a sharp turn that favored AMD: sub-$1,000 personal computers had become the fastest growing part of the market, and AMD's K6 chips were well-suited to this shift. Here again, AMD had a perfect scenario. Customers wanted to lessen Intel's power, the market shift toward cheaper computers gave AMD an edge, and the K6 was an ideal product at exactly the right moment in the midst of one of the greatest technology booms in history.[33]

And then . . . AMD failed to execute brilliantly, unable to make enough chips to meet demand. Customers rooted for AMD—they really *wanted* a viable alternative to Intel—but they couldn't reliably get enough of the K6 due to AMD's manufacturing problems, and they began to turn back to Intel. Despite an extraordinary run of good luck at the best possible moment, AMD's stock fell more than 70 percent behind the general stock market from the start of 1995 through the end of 2002.[34]

> The AMD story illustrates a common pattern we observed in the comparison companies during their respective eras of analysis, the squandering of good luck. When the time came to execute on their good fortune, they stumbled. They didn't fail for lack of good luck; they failed for lack of superb execution.

In 1980, IBM sought an operating system for its then-in-development personal computer. We now know that this led to a turning point in

Microsoft's history, but when IBM first went looking, the outcome could've been very different. Microsoft didn't have an operating system or even have plans to be in the operating-system business. The clear front-runner, the company that should have established the dominant standard in the personal computer business, was Digital Research in Pacific Grove, California. Digital Research would have been a comparison candidate in our research but was excluded due to being privately held; still, the story is worth sharing to highlight the question, "When the moment comes, will you capture it, or just let it slip?"

Digital Research had developed CP/M, the leading non-Apple operating system for personal computers, and IBM executives traveled to Digital Research's offices to discuss the possibility of working together. Digital Research's CEO, Gary Kildall, had a previously scheduled business meeting in the Bay Area, and piloting his own private plane, he flew up to San Francisco, leaving the first part of the IBM meeting in the hands of colleagues. By the time Kildall piloted himself back in the afternoon, the meeting had taken a negative turn. The IBM people left later that day, unimpressed, and Kildall departed for a vacation. Accounts vary as to precisely why the talks disintegrated, but the result was that IBM turned to Microsoft in frustration.[35] Microsoft recognized the moment and committed itself to a brutal schedule to get an operating system ready for the launch of the IBM PC.[36] Digital Research had the incredibly good fortune to be in the right place at the right time when IBM came knocking, but it didn't get a great return on luck. Microsoft did.

10XERS SHINE: GREAT RETURN ON BAD LUCK

On November 8, 1988, Peter Lewis received news that shocked and stunned the insurance industry. California voters passed Proposition 103, a punitive attack on car-insurance companies, mandating 20 percent price reductions and refunds to customers, and plunging the world's largest auto-insurance market into chaos. Progressive Insurance had significant exposure, with nearly a quarter of its entire business from that one state—bang!—severely damaged by a 51 percent vote on a single day.[37]

Lewis *zoomed out* to ask, "*What the heck is going on?*" He placed a call to his former Princeton classmate, Ralph Nader. Nader had long been a consumer-rights activist, at one point leading a sort of special-forces unit nicknamed Nader's Raiders, and he'd championed Proposition 103. The message Lewis heard: *People hate you.* People simply hated dealing with insurance companies and they revolted, screaming with their votes. "People were saying 'We hate your guts, we're going to kill you and we don't give a damn,' " said Lewis. Chastened by what he'd heard, Lewis called his staff together, told everyone, "Our customers actually hate us," and challenged his team to create a better company.[38]

Lewis came to see Proposition 103 as a gift, and he used this gift to deepen the company's core purpose, to reduce the economic cost and trauma caused by auto accidents. So, Progressive created "Immediate Response" claims service. No matter what time you had an accident—24 hours a day, seven days a week, 365 days a year—Progressive would be available to help. Claims adjusters would work from a fleet of vans and SUVs dispatched to policyholder homes or even directly to an accident. By 1995, 80 percent of the time the Progressive adjuster would've gone to a customer ready to issue a check within 24 hours of an accident. In 1987, the year before Proposition 103, Progressive ranked #13 in the American private-passenger auto-insurance market; by 2002, it reached #4. Years later, Peter Lewis called Proposition 103 "the best thing that ever happened to this company."[39]

> Progressive and Peter Lewis illustrate how 10Xers shine when clobbered by setbacks and misfortune, turning bad luck into good results. 10Xers use difficulty as a catalyst to deepen purpose, recommit to values, increase discipline, respond with creativity, and heighten productive paranoia. Resilience, not luck, is the signature of greatness.

As we were working on this research, we read about an analysis of Canadian-born hockey players, wherein academic researchers identified a correlation between birth date and hockey success. Those born in the second half of the year had less success than those born in the first half of the year. Being 10¾ years old versus 10 years old can make a difference in terms of size and speed. So, with an age-class cutoff of January 1, the kids born at the beginning of the year have a physical advantage over those born at the end of the year, which then compounds as they have more early success and garner more attention from coaches. Author Malcolm Gladwell popularized these findings, writing that this pattern eventually played out all the way to the National Hockey League (NHL), where the distribution of birth dates is skewed to the first half of the year by 70 percent to 30 percent.[40]

But a closer look at the data leads to a very different conclusion for truly great hockey players, the 10Xers, those few who make it to the Hockey Hall of Fame. (Those who make it to the Hall of Fame are members of a much more elite group than those who only make it to the NHL. The Hall of Fame currently inducts no more than four players per year, and induction is based on a player's entire career.) In fact, half of Canadian-born Hall of Famers had birth dates in the *second* half of the year. (See *Research Foundations: Hockey Hall of Fame Analysis.*) Now, consider the following. If indeed a substantially lower percentage of Canadian-born NHL players are born in the second half of the year than in the first half of the year, yet half of Canadian-born Hall of Fame inductees have birth dates in the second half of the year, this leads to a very interesting inversion: Canadian NHL players with the "bad luck" of being born in the second half of the year have a *higher* likelihood of making it into the Hall of Fame than those with the "good luck" of being born in the first half of the year![41]

Consider Ray Bourque, born in December, who came from a poor family, grew up in a working-class neighborhood, lived in an apartment with children "stacked from floor to ceiling in bunk beds," and thrilled at even having skates at all. Bourque lived hockey, sleeping with his skates, creating a makeshift rink in the cellar of his apartment building, practicing thousands of shots, blasting the puck at a goal pinned to the wall so hard that it cracked the cement, water leaking in, his father repairing the dingy walls with crack filler. Bourque developed a crushing work ethic that endured; for most of his NHL career en route to the Hall of Fame, he played more than thirty minutes a game, at times double that of his teammates, reflecting his prodigious, self-imposed fitness regimen. He played in 19 consecutive NHL All-Star games and retired as the most proficient scoring defenseman in NHL history. Bourque was a gifted physical specimen, and he likely had superior skills even as a youngster. But most players who make it to the NHL are also impressive physical specimens, and most likely had outstanding skills even as youngsters. There are far fewer, however, who prove themselves to be 10Xers across an entire career, like Ray Bourque.[42] "Goals live on the other side of obstacles and challenges," said Bourque. "Along the way, make no excuses and place no blame."[43] Bourque had luck in his journey, good and bad, but luck did not *make* Bourque into one of the greatest hockey players of all time.

Now, you might be thinking, "But Bourque is an exception."

Precisely. The whole point *is* to become exceptional.

Nietzsche famously wrote, "What does not kill me, makes me stronger."[44] We all get bad luck. The question is how to use that bad luck to make us stronger, to turn it into "one of the best things that ever happened," to not let it become a psychological prison. And that's precisely what 10Xers do.

BAD LUCK, POOR RETURN: THE ONE PLACE YOU REALLY DON'T WANT TO BE

We came across a remarkable moment at the very start of Southwest Airlines' life, described by its first CEO, Lamar Muse, in his book, *Southwest Passage*: "The very first Sunday morning of Southwest's life, we narrowly escaped a disaster . . . During the takeoff run, the right thrust-reverser deployed. Only the captain's instantaneous reaction allowed him to recover control and make a tight turn for an emergency landing on one engine."[45] What if, despite the pilot's heroics, he'd not been able to stop the aircraft from a flat spin? What if the 737 had smashed into the ground in the first week of building the brand? Would there even be a Southwest Airlines today?

There's only one truly definitive form of luck, and that's the luck that ends the game. If Southwest missed an opportunity to open in a new city or grab a set of gates at a new airport, it still could have turned itself into a great company. But if Southwest had been knocked out of business with a plane crash in its first week of operation, it likely would have lost forever the chance to become a great company. Recall the essential first half of Nietzsche's quote, "*What does not kill me . . .*"

> There's an interesting asymmetry between good luck and bad luck. A single stroke of good luck, no matter how big the break, cannot by itself make a great company. But a single stroke of extremely bad luck that slams you on the Death Line, or an extended sequence of bad-luck events that creates a catastrophic outcome, can terminate the quest.

In the late 1970s and early 1980s, both PSA and Southwest struggled with a similar sequence of bad-luck events. *Both* companies got smacked by an oil shock that spiked jet-fuel prices; *both* companies experienced an air-traffic-control strike; *both* companies faced a severe recession and spiraling inflation (particularly difficult for airlines); *both* companies suffered from skyrocketing interest rates that increased the cost of jet leasing; *both* companies had an unexpected change of CEOs. As PSA's President Paul Barkley noted in 1982, "It has been less than two years . . . it seems more like ten years have gone by."[46] From 1979 through 1985, PSA fell into a self-destructive Doom Loop, raising prices rather than lowering costs, destroying its culture with layoffs and acrimonious labor battles, downgrading its balance sheet with increasing debt, and putting in place a CEO who abandoned the SMaC recipe and delivered erratic earnings. PSA got poor returns on bad luck and fell permanently behind Southwest.[47]

If we all get some combination of both heads (lucky flips) and tails (unlucky flips), and if the ratio of heads to tails tends to even out over time, we need to be skilled, strong, prepared, and resilient to endure the bad luck long enough to eventually get good luck. Malcolm Daly had to be lucky enough to survive the fall, but he also had to be strong, skilled, and resilient *before* the 44 hours of peril after his two-hundred-foot fall. The Southwest pilot had to be skilled and prepared *before* the thrust-reverser deployed, and the Southwest spirit had to be strong and resilient *before* the bad luck of the early 1980s.

As we discussed in Chapter 5, 10Xers exercise productive paranoia, combined with empirical creativity and fanatic discipline, to create huge margins of safety. If you stay in the game long enough, good luck tends to return, but if you get knocked out, you'll never have the chance to be lucky again. Luck favors the persistent, but you can persist only if you survive.

Dane Miller grasped this idea in the early days of 10X-case Biomet, running lean to the extreme to buffer against whatever the company might encounter in its fledgling years from 1977 to 1982. Miller and three colleagues quit their jobs and threw their personal savings into the company, working 12 to 16 hours a day (including weekends) in a ramshackle space—a converted barn, actually—with a hole cut in the wall to attach a mobile home for storing inventory. They'd leave the air-conditioning off as long as possible in the summer to minimize utility charges, people working at fold-up card tables with beads of sweat dripping off their noses. To save money on a financing trip, Miller and one of his colleagues spent the night in the motor home of a Presbyterian church and had to shower in ice-cold water. At one point, Miller noticed an empty field behind their headquarters, and he had an idea: Why not raise cows, letting them graze on the unused grassland? If the company ran out of cash, they could eat the cows to get through a rough patch. So, they herded three cows onto the lot, making Biomet the first cattle-farming hedge play in the medical-devices industry.

Biomet had to endure more than five tough years before it obtained substantial outside funding, trying a range of product possibilities, eating the cows, and taking cold showers along the way. It survived being turned down by venture capitalists. It survived when its subcontract manufacturers failed to deliver parts Biomet needed. It survived being turned down by established distributors. It survived long enough for its implant products to finally gain traction, setting the company on a path to beat its industry by more than 11 times.[48]

LUCK IS NOT A STRATEGY

Life offers no guarantees. But it does offer strategies for managing the odds, indeed, even managing luck. The essence of "managing luck" involves four things: (1) cultivating the ability to *zoom out* to recognize luck when it happens, (2) developing the wisdom to see when, and when not, to let luck disrupt your plans, (3) being sufficiently well-prepared to endure an inevitable spate of bad luck, and (4) creating a positive return on luck—both good luck and bad—when it comes. Luck is not a strategy, but getting a positive return on luck is.

And how would you get the highest possible return on luck? It turns out that you've been reading about it all the way along in the previous chapters. Keep in mind the original premise of the study: life is uncertain, full of big, consequential forces that we can neither predict nor control. Luck is uncertain, uncontrollable, and consequential. Indeed, we could reframe the entire study around luck and how to get a great ROL.

Let's review where we've been:

10Xer behaviors: Leaders with fanatic discipline, empirical creativity, productive paranoia, and Level 5 ambition never relax when blessed with good luck. They never wallow in despair when hit with bad luck. They keep pushing, driving for the overall goal and cause.

20 Mile March: When 10Xers get a lucky break, they seize it and then build upon it, not just for days or weeks but for years or decades. A 10Xer builds a culture that can achieve results whether it gets good luck or bad, engendering deep confidence that success, in the end, doesn't depend upon luck.

Fire bullets, then cannonballs: While 10Xers don't "cause" their luck, they increase the chances of stumbling upon something that works by firing lots of bullets. By marrying creativity with empirical validation, 10Xers can fire big cannonballs that don't rely on luck for ultimate success. Uncalibrated cannonballs require luck for a successful outcome; calibrated cannonballs do not.

Leading above the Death Line: By having lots of extra oxygen canisters (building big buffers and margins of safety), 10Xers give themselves more options for responding to luck. By managing three types of risk—Death Line risk, asymmetric risk, and uncontrollable risk—they shrink the odds of catastrophe in the face of bad luck. The ability to *zoom out,* then *zoom in* helps them recognize luck and consider if it merits disrupting their plans.

SMaC: SMaC behaviors minimize mistakes that can amplify bad-luck events. They also increase the odds of executing brilliantly when a good-luck moment arrives. Having a clear SMaC recipe can help you decide whether and how to let a luck event disrupt your plans.

All the concepts in this book contribute to getting a high ROL. 10Xers recognize that we're all swimming in a sea of luck. They understand that we cannot cause, control, or predict luck. But by behaving and leading in 10X ways, they make the most of the luck they get. There's an adage that says "Better to be lucky than good." And it's perhaps true—for those who seek to be only good, not much better than average, creating nothing exceptional. But our research brings us to an entirely opposite conclusion for those who aspire to more: it's far *better to be great than lucky.*

The best leaders we've studied maintain a paradoxical relationship to luck. On the one hand, they credit good luck *in retrospect* for having played a role in their achievements, despite the undeniable fact that others were just as lucky. On the other hand, they don't blame bad luck for failures, and they hold only themselves responsible if they fail to turn their luck into great results. 10Xers grasp that if they blame bad luck for failure, they capitulate to fate. Equally, they grasp that if they fail to perceive when good luck helped, they might overestimate their own skill and leave themselves exposed when good luck runs dry. There might be more good luck down the road, but 10Xers never count on it.

CHAPTER SUMMARY

RETURN ON LUCK

KEY POINTS

▶ We defined a luck event as one that meets three tests: (1) some significant aspect of the event occurs largely or entirely independent of the actions of the key actors in the enterprise, (2) the event has a potentially significant consequence (good or bad), and (3) the event has some element of unpredictability.

▶ Luck happens, a lot, both good luck and bad luck. Every company in our research experienced significant luck events in our era of analysis. Yet the 10X cases were *not* generally luckier than the comparison cases.

- The 10X companies did not generally get more good luck than the comparisons.
- The 10X companies did not generally get less bad luck than the comparisons.
- The 10X companies did not get their good luck earlier than the comparisons.
- The 10X companies cannot be explained by a single giant-luck spike.

▶ We've encountered two extreme views on the topic of luck. One extreme holds that luck is the primary cause of 10X success; the other extreme holds that luck plays no role in 10X success. Both views are not supported by the evidence from our research. The critical question is not "Are you lucky?" but "Do you get a high *return on luck*?"

CHAPTER SUMMARY

▶ There are four possible ROL scenarios:

- Great return on good luck
- Poor return on good luck
- Great return on bad luck
- Poor return on bad luck

▶ We observed an asymmetry between good luck and bad. A single stroke of good fortune, no matter how big, cannot by itself make a great company. But a single stroke of extremely bad luck, or an extended sequence of bad-luck events that create a catastrophic outcome, can terminate the quest. There's only one truly definitive form of luck, and that's the luck that ends the game. 10Xers assume they'll get a spate of bad luck and prepare ahead of time.

▶ The leadership concepts in this book—*fanatic discipline*; *empirical creativity*; *productive paranoia*; *Level 5 ambition*; *20 Mile March*; *fire bullets, then cannonballs*; *leading above the Death Line*; and *SMaC*—all contribute directly to earning a great ROL.

▶ 10Xers credit good luck as a contributor to their success, despite the undeniable fact that others also experienced good luck, but they never blame bad luck for setbacks or failures.

UNEXPECTED FINDINGS

▶ Some of the comparison companies had extraordinarily good luck, better luck even than the 10X winners, yet failed because they squandered it.

▶ 10X cases got a substantial amount of bad luck yet managed to get a great ROL. This is when 10Xers really shine, exemplifying the philosophy, "What does not kill me, makes me stronger."

▶ ROL might be an even more important concept than return on assets (ROA), return on equity (ROE), return on sales (ROS), or return on investment (ROI).

▶ "Who Luck"—the luck of finding the right mentor, partner, teammate, leader, friend—is one of the most important types of luck. The best way to find a strong current of good luck is to swim with great people, and to build deep and enduring relationships with people for whom you'd risk your life and who'd risk their lives for you.

KEY QUESTIONS

▶ What significant luck events have you experienced in the last decade? Did you get a high return on luck? Why or why not? What can you do to increase your return on luck?

BONUS QUESTION

▶ Who is your best luck?

EPILOGUE

GREAT BY CHOICE

"One should . . . be able to see that things are hopeless
and yet be determined to make them otherwise."

—F. Scott Fitzgerald[1]

We sense a dangerous disease infecting our modern culture and eroding hope: an increasingly prevalent view that greatness owes more to circumstance, even luck, than to action and discipline—that what happens to us matters more than what we do. In games of chance, like a lottery or roulette, this view seems plausible. But taken as an entire philosophy, applied more broadly to human endeavor, it's a deeply debilitating life perspective, one that we can't imagine wanting to teach young people. Do we really believe that our actions count for little, that those who create something great are merely lucky, that our circumstances imprison us? Do we want to build a society and culture that encourage us to believe that we aren't responsible for our choices and accountable for our performance?

Our research evidence stands firmly against this view. This work began with the premise that most of what we face lies beyond our control, that life is uncertain and the future unknown. And as we wrote in Chapter 7, luck plays a role for everyone, both good luck and bad luck. But if one company becomes great while another *in similar circumstances and with comparable luck* does not, the root cause of why one becomes great and the other does not simply cannot be circumstance or luck. Indeed, if there's one overarching message arising from more than six thousand years of corporate history across all our research—studies that employ comparisons of great versus good in similar circumstances—it would be this: greatness is not primarily a matter of circumstance; greatness is first and foremost a matter of conscious choice and discipline. The factors that determine whether or not a company becomes truly great, even in a chaotic and uncertain world, lie largely within the hands of its people. It is not mainly a matter of what happens *to* them but a matter of what they create, what they do, and how well they do it.

This book and the three that precede it (*Built to Last, Good to Great,* and *How the Mighty Fall*) are looks into the question of what it takes to build an enduring great organization. As we conducted the 10X research, we simultaneously tested the concepts from the previous work, considering whether any of the key concepts from those works ceased to apply in highly uncertain and chaotic environments. The earlier concepts held up, and we are confident that the concepts from all four studies increase the odds of building a great company.

But do they *guarantee* success? No, they don't. Good research advances understanding but never provides the ultimate answer; we always have more to learn. And life offers no guarantees. It's always possible that game-ending events and unbendable forces—disease, accident, brain injury, earthquake, tsunami, financial calamity, civil war, or any of a thousand other possible events—will subvert our strongest and most disciplined efforts. Still, we must act.

When the moment comes—when we're afraid, exhausted, or tempted—what choice do we make? Do we abandon our values? Do we give in? Do we accept average performance because that's what most everyone else accepts? Do we capitulate to the pressure of the moment? Do we give up on our dreams when we've been slammed by brutal facts? The greatest leaders we've studied throughout all our research cared as much about values as victory, as much about purpose as profit, as much about being useful as being successful. Their drive and standards are ultimately internal, rising from somewhere deep inside.

We are not imprisoned by our circumstances. We are not imprisoned by the luck we get or the inherent unfairness of life. We are not imprisoned by crushing setbacks, self-inflicted mistakes or our past success. We are not imprisoned by the times in which we live, by the number of hours in a day or even the number of hours we're granted in our very short lives. In the end, we can control only a tiny sliver of what happens to us. But even so, we are free to choose, free to become great by choice.

FREQUENTLY ASKED QUESTIONS

Q: Were any of the concepts from *Good to Great, Built to Last,* or *How the Mighty Fall* overturned by this research?

No. As we conducted the 10X research, we systematically examined the relationship of the 10X cases (and their comparisons) to concepts from the prior work. The evidence showed that the 10X cases exemplified the prior concepts to a greater degree than the comparison cases.

Q: To what extent did the Level 5 leaders in *Good to Great* exhibit 10Xer behaviors?

We observed fanatic discipline, empirical creativity, and Level 5 ambition in the Level 5 leaders in the *Good to Great* research, very much as with the 10Xers; however, we observed less productive paranoia in the good-to-great leaders than in the 10X leaders in this study. We believe this is because they operated in less severe environments. Recall the analogy from Chapter 1 about going on a leisurely hike, with warm, sunlit meadows on a warm spring day with a truly great mountaineering expedition leader. In those situations, you wouldn't see everything that makes him different from others. The Level 5 leaders in *Good to Great* operated in safer environments than the 10Xers. Also, the good-to-great leaders generally took over already-established (and often quite large) good companies, whereas the 10Xers in this study began as entrepreneurs or small-business leaders, which rendered them more exposed and vulnerable to their environments. If Level 5s in *Good to Great* had been leading small companies facing the level of uncertainty and chaos faced by the 10X leaders in this study, we suspect they would have shown more productive paranoia. Finally, we'd note that *Good to Great* perhaps put more emphasis on the humility aspect of the Level 5 duality (a Level 5 leads with a paradoxical blend of personal humility and professional will), whereas this study highlights more of the will aspect. To be a true Level 5 leader, however, always requires exercising both humility *and* will.

Q: What role does the "First Who" principle play when leading a company amidst uncertainty and chaos (the concept of getting the right people on the bus, the wrong people off the bus, and the right people into key seats; and then figuring out where to drive the bus)?

We didn't write much about First Who in this book because the concept is so heavily covered in *Good to Great*. But make no mistake: 10X leaders are fanatic about getting the right people on the bus and into the right seats. Recall David Breashears's dedication to having the right people on Everest, living by the adage that a summit team is only as strong as its weakest member. *Time* magazine wrote of Southwest Airlines in 2002, "The airline received 200,000 resumes last year but hired only 6,000 workers—making it more selective than Harvard." Progressive Insurance identified having the right people as the #1 strategic pillar for accomplishing its objectives and beating its competition, noting proudly in 1990 that "there are 15 people who we asked to leave who became presidents of other insurance companies." John Brown at Stryker had a gift for picking the right people and the discipline to move people out of seats in which they were failing, following a Stryker philosophy that it's better to invest heavily in the right people than to pour too much energy into people who aren't going to make it. George Rathmann said of Amgen's early history, "Amgen is one of those companies where all the assets go home at night in tennis shoes," and by the 1990s, Amgen rejected 57 of every 58 job applicants. Intel cofounder Robert Noyce assembled Intel's founding team *before* deciding what products to make; he took personal responsibility for recruiting Intel's early talent and believed that the right people in the right culture would lead to great outcomes. As Tom Wolfe wrote about Ted Hoff and his invention of the microprocessor: "Noyce took Hoff's triumph as proof [that if] you created the right type of corporate community, the right type of autonomous congregation, genius would flower." Microsoft used extreme standards to select the right people for Microsoft, with Gates's summing up in 1992, "Take away our 20 best people and I tell you that Microsoft would become an unimportant company." Biomet paid fastidious attention to getting the right people in every seat, using stock options at all levels to attract and retain the best talent.[1]

All the 10X companies cultivated cult-like cultures wherein the right people would flourish and equally, where the wrong people would quickly self-eject. The 10X study is predicated on the premise of unending uncertainty, which increases the importance of First Who; if you cannot predict what's going to happen, you need people on the bus who can respond and adapt successfully to whatever unforeseen events might hit.

Q: Is there a relationship between the SMaC recipe and the Hedgehog Concept from *Good to Great*?

A Hedgehog Concept is a simple, crystalline concept that flows from a deep understanding about the intersection of the following three circles:

1. What you are passionate about
2. What you can be the best in the world at
3. What drives your resource or economic engine

Once the good-to-great companies were clear on their Hedgehog Concepts, they built momentum by making a series of decisions relentlessly consistent with that concept, like turning a giant, heavy flywheel, turn upon turn. A SMaC recipe is the code for translating a high-level Hedgehog Concept into specific action and for keeping an organization focused in the same direction, thereby building flywheel momentum. (See adjacent diagram that shows Hedgehog Concept, SMaC Recipe, and Flywheel Effect.) Southwest Airlines, for example, had a high-level Hedgehog Concept: to be the best high-spirit, low-cost airline, steadily increasing profit-per-fuselage, with great passion for being an industry renegade. It translated this high-level concept into Putnam's 10 points, discussed in Chapter 6. By consistently adhering to the recipe, Southwest built cumulative momentum in the flywheel, flight by flight, city by city, gate by gate, year by year, to rise from a start-up in Texas to become the most successful airline.

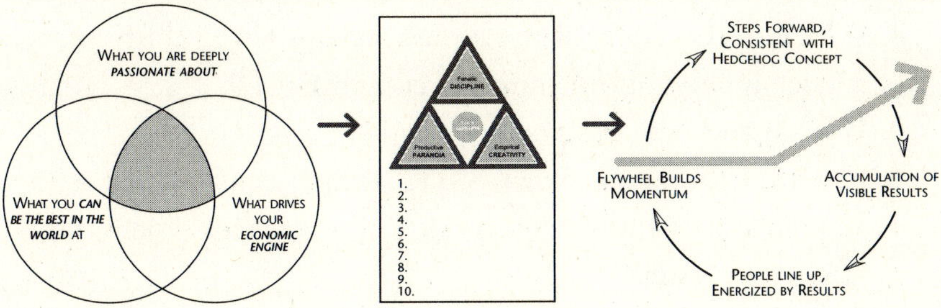

HEDGEHOG CONCEPT SMAC RECIPE FLYWHEEL EFFECT

Q: Do you have any guidance for how to craft a SMaC recipe?

The key to crafting a SMaC recipe is to go directly to the practical, the empirical, and when possible, the specific and concrete. You can vaguely aspire to "high aircraft utilization," or like Southwest Airlines, you can specify "gate turns of 10 minutes" or "fly only 737s." You can aim without precision to "advance technology," or like Intel, you can focus on a more concrete task: "double the number of components every two years." You can seek to "be efficient with the camera" or you can specify, "Be able to assemble the camera, mount on tripod, load and thread film, aim and shoot in five minutes flat."

The SMaC recipe should reflect insight—based on empirical validation—about what works, and why. It should help make it clear what to do and what not to do. It should be durable, so that it requires only amendments, not wholesale revolution, in response to changing conditions. When formulating a SMaC recipe, ask, "What durable and specific practices best drive our results?" In laboratory working sessions with executives, we've employed the following methodology:

1. Make a list of successes your enterprise has achieved.
2. Make a list of disappointments your enterprise has experienced.
3. What *specific* practices correlate with the successes but *not* the disappointments?
4. What *specific* practices correlate with the disappointments but *not* the successes?
5. Which of these practices can last perhaps ten to thirty years and apply across a wide range of circumstances?
6. *Why* do these specific practices work?
7. Based on the above, what SMaC recipe, consisting of 8 to 12 points that reinforce each other as a coherent system, best drives your results?

Q: If the 10X concepts are universal, why didn't they become starkly clear in *Good to Great*?

As we wrote in Chapter 1, each research study is like poking holes in the side of a black box and shining a light inside to see the inner workings of the principles that make great companies. Each hole provides a different perspective. The *Good to Great* study focused on how to make a leap from oppressive mediocrity to great results. We selected the good-to-great companies based on a pattern of 15 years of mediocre performance punctuated by a breakthrough to 15 years of exceptional performance, not on the severity of the environment. This study, in contrast, looked through an entirely different hole punched in the black box, selecting small or start-up companies that became great in uncertain, unforgiving, and chaotic environments. There's no inconsistency between the studies or their findings, just very different angles of analysis. The two studies don't repeat each other, nor do they contradict each other; they complement each other.

Q: If I'm not a full 10Xer, can I compensate by building a 10X team that has all the behaviors?

Instead of focusing on whether any given individual is a 10Xer, it's better to focus on working as a team to implement the key ideas in Chapters 3 through 7 as an entire enterprise. Set a 20 Mile March and commit to it. Fire bullets, then fire calibrated cannonballs. Practice all the elements of productive paranoia discussed in "Leading above the Death Line." Adhere to and selectively amend a SMaC recipe. Become highly attuned to luck, and respond to every luck event, good or bad, with the question, "What are we going to do to get a high return on this luck (ROL)?" If your team and enterprise succeed at all of these, it will matter less whether any single individual is a full-fledged 10Xer.

Q: Does leading above the Death Line mean avoiding BHAGs (Big Hairy Audacious Goals)?

No. Roald Amundsen en route to the South Pole and David Breashears with his IMAX camera on Everest were pursuing BHAGs, as were the 10X leaders in our research-study companies. The task is to pursue BHAGs *and* stay above the Death Line.

Q: How is the 10X concept "fire bullets, then cannonballs" different from the *Built to Last* concept "try a lot of stuff and keep what works"?

The two ideas overlap, but the key additional insight from the 10X research is that 10Xers follow up successful bullets with cannonballs. Trying a lot of stuff is, in essence, firing bullets. But keeping what works is not the same thing as making a big bet to fully exploit what you've learned from firing a bullet. That's what cannonballs are for.

Q: What are the implications for innovation-driven economies of your finding that 10X cases didn't always out-innovate comparison companies?

Our research suggests that treating innovation *alone* as the silver bullet for achieving a competitive advantage would be naïve and unwise. We conclude that 10X success requires the ability to *scale* innovation with great consistency, by blending creativity and discipline to build organizations that turn innovation into sustained great performance. This is the Intel story. It's also the Southwest story, the Microsoft story, the Amgen story, the Stryker story, the Biomet story, the Progressive story, the story of the resurgence of Genentech under Levinson, and even the Apple story during its best years. If an enterprise—whether a company or a nation—retains its creativity yet loses discipline, increases pioneering innovation yet forgets how to multiply that innovation at scale (and at minimum cost), our research suggests that enterprise will be at risk.

Q: You mention the "Genius of the AND" a few times in the text. What's the Genius of the AND and how does it apply here?

We found in the *Built to Last* study that leaders of enduring great companies are comfortable with paradox, having the ability to embrace two opposed ideas in the mind at the same time. They don't oppress themselves with what we call the "Tyranny of the OR," which pushes people to believe that things must be either A OR B, but not both. Instead, the best leaders liberate themselves with the Genius of the AND—the ability to embrace both extremes of a number of dimensions at the same time. In the words of F. Scott Fitzgerald, "The test of a first-rate intelligence is the ability to hold two opposed ideas in the mind at the same time, and still retain the ability to function." In the 10X study, we found extensive evidence of the Genius of the AND. For example,

Disciplined	*And*	Creative
Empirical validation	*And*	Bold moves
Prudence	*And*	BHAGs (Big Hairy Audacious Goals)
Paranoid	*And*	Courageous
Ferociously ambitious	*And*	Not egocentric
Severe performance standards, no excuses	*And*	Never going too far, able to hold back
On a 20 Mile March	*And*	Fire bullets, then cannonballs

Threshold innovation	*And*	One fad behind
Cannot predict the future	*And*	Prepared for what they cannot predict
Go slow when they can	*And*	Go fast when they must
Disciplined thought	*And*	Decisive action
Zoom out	*And*	*Zoom in*
Adhering to a SMaC recipe	*And*	Amending a SMaC recipe
Consistency	*And*	Change
Never count on luck	*And*	Get a high ROL when luck comes

Q: How do you respond to critics of your research findings who point to the failings of previously great companies you've researched and written about?

As we discussed in Chapter 1, our research is based upon studying specific, dynastic eras of performance, like studying the greatest sports dynasties in history. That some sports dynasties later cease to be dynasties would be irrelevant to the overall analysis of what it takes to build a great sports dynasty.

Q: Can this book help companies avoid the five stages of decline outlined in *How the Mighty Fall*?

Yes. In fact, the comparison cases in this study that fell from potential greatness (PSA, Safeco, USSC, Genentech pre-Levinson, and Apple before the return of Steve Jobs) all showed elements of Stages 1 through 4 of decline, and some went all the way to Stage 5. (See diagram, "The Five Stages of Decline.") The 10X concepts in this work can play a significant role in staving off the stages of decline. Doing a 20 Mile March, avoiding uncalibrated cannonballs, and adhering to a SMaC recipe help companies stay out of Stage 2. "Leading above the Death Line" concepts (amassing oxygen canisters, bounding risk, and *zooming out/zooming in*) directly aid in keeping Stage 3 at bay. Carefully amending a SMaC recipe (rather than inciting wholesale, reactive revolution) enables companies to avoid Stage 4. As for the peril of Stage 1, hubris, those who truly practice productive paranoia never feel they're invincible; they always fear that potential doom lurks just around the corner.

The Five Stages of Decline from ***How the Mighty Fall*** by Jim Collins

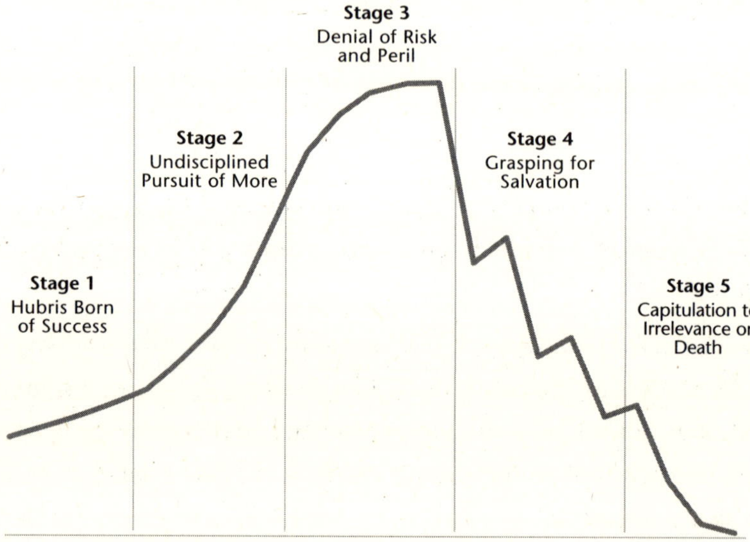

Q: How did you two (Jim and Morten) begin your working partnership, and why did you do this research project as a team?

We first met at Stanford Graduate School of Business in 1991. Jim, then teaching entrepreneurship and small business, and his colleague Professor Jerry Porras had embarked on the *Built to Last* research project, and Morten joined the research team en route to receiving his PhD. Later, while a faculty member at Harvard Business School, Morten contributed critical input on research methods and study design for Jim's *Good to Great* project. We always talked about collaborating on a project from the ground up if we discovered a mutually fascinating question. The question behind this book—why do some thrive in the face of immense uncertainty, even chaos, and others do not?—had been gestating in our minds for years, but had been pushed to the background while working on other projects. Then, in the aftermath of 9/11 and the bursting stock bubble, watching the exponential rise of global competition and the relentless onslaught of technological disruption, hearing the rising chant of "change, change, change," the question asserted itself. We both came to believe that uncertainty is permanent, chaotic times are normal, change is accelerating, and instability will likely characterize the rest of our lives.

Q: Do you see your book as about defining and thriving in a New Normal?

No. The premise behind this work is that instability is chronic, uncertainty is permanent, change is accelerating, disruption is common, and we can neither predict nor govern events. We believe there will be no "new normal." There will only be a continuous series of "not normal" times.

The dominant pattern of history isn't stability, but instability and disruption. Those of us who came of age amidst stable prosperity in developed economies in the second half of the 20th century would be wise to recognize that we grew up in a historical aberration. How many times in history do people operate inside a seemingly safe cocoon, during an era of relative peace, while riding one of the most sustained economic booms of all time? For those of us who grew up in such environments—and especially for those who grew up in the United States—nearly all our personal experience lies within a rarified slice of overall human history, very unlikely to repeat itself in the 21st century and beyond.

Q: How widely applicable is the question underlying this study? Do you see it as universal?

Stop to think about your own situation or organization, and consider the following question. Rate the context in which you operate today on a 1-to-10 scale. A rating of 1 means you face no big forces outside your control, nothing moves particularly fast, you can predict most of what's going to happen, everything feels stable and certain, and there's nothing out there that can significantly alter your trajectory (good or bad). A rating of 10 means you face tremendous, fast-moving, unpredictable forces outside of your control, that elicit feelings of uncertainty and instability, and that can have a huge impact (good or bad) on your trajectory. How would you rate your environment—stable or unstable, certain or uncertain, predictable or unpredictable, in your control or not, more like a 3 or more like an 8?

It doesn't matter whether we're discussing this question with small-company entrepreneurs, Army generals, K–12 educators, church leaders, membership associations, police chiefs, city managers, healthcare professionals, philanthropists, CIOs, CFOs, CEOs, or even individuals concerned about their jobs and families. When we ask this question, we get a remarkably consistent pattern of answers. After giving people a moment to reflect, we ask for a show of hands.

"How many have a score of less than 5?"

Almost no hands go up.

"How many have a score of 5 or 6?"

A few hands go up.

"How many have a score of 7 or 8?"

More than half the people in the room raise their hands.

"How many have a score of 9 or 10?"

The remaining people raise their hands.

The question of what it takes to thrive in the face of uncertainty, even chaos, feels relevant to every industry and every social sector we've encountered so far.

Q: Do you see the causes of chaos and uncertainty as primarily economic?

Not entirely. Certainly, there are economic drivers, such as increased global competition, volatile capital markets, and rapidly evolving business models. But clearly, the sources of instability come from far outside economics, such as government regulation (or deregulation), undisciplined government spending, unpredictable political risk, disruptive technologies, new media, the amplifying effect of a 24-hour news cycle, natural disasters, terrorism, energy shocks, climate change, political upheaval in emerging countries, and so on. And there'll be entirely new disruptions and chaotic forces as yet unforeseen.

Q: Do you see this book as about the past or the future?

We've studied the past, but we see this book as having great relevance for leading in the future. Our strategy was to carefully examine companies that had achieved greatness in the most uncertain and chaotic industries, and to glean the general principles for thriving in such environments so that they can be applied by all enterprises dealing with the uncertainty and episodes of chaos in the 21st century.

Q: My world feels fairly stable right now; does this apply to me?

Remember a lesson from Chapter 5: it's what you do *before* the storm comes that most determines how well you'll do *when* the storm comes. Those who fail to plan and prepare for instability, disruption, and chaos in advance tend to suffer more when their environments shift from stability to turbulence.

Q: Do the 10X concepts apply as much to the social sectors as the *Good to Great* ideas?

While conducting this research, we simultaneously worked with leaders from a wide range of social sectors, including K–12 education, higher education, churches, nonprofit hospitals, the military, police forces, government (city, county, state, and national), museums, orchestras, social-safety-net (hunger- and homelessness-related) organizations, youth programs, and a wide range of cause-driven nonprofits. Like business leaders, they face big forces outside their control; high degrees of uncertainty; fast-moving events; dangerous threats; and huge, disruptive opportunities. We've found these ideas to be directly relevant for them, albeit with unique translation to each sector.

Q: Do you see this work as being primarily about navigating in times of austerity and crisis?

No. This is *not* a book on crisis management, nor is it about thriving amidst recession or even economic calamity. Crises and "difficult times" are simply special-case scenarios of a more general condition of unrelenting instability and chronic uncertainty, whether in good times or bad. In fact, disruptive opportunity is just as dangerous as disruptive threat. Times of explosive growth are at least as difficult to navigate as times of economic austerity.

Keep in mind some of the industries we studied: software, computers, microelectronics, biotechnology, insurance, and medical devices. These industries were full of spectacular growth and opportunity, while also being uncertain and chaotic. Consider computer software. In 1983, *Industry Week* magazine published a story entitled "Software Sparks a Gold Rush" and listed the top 16 personal computer–software companies. All 16 sat right on the nose cone of a rocket about to take off, a nascent industry that would sell more than a billion personal computers worldwide by the early 2000s. Yet along the way, most of the early leaders lost their independence, and some died outright. Of the 16 leaders listed in the 1983 article, only 3 remain standing as independent companies at the time of this writing. The opportunity was huge, the amount of change was huge, and the resulting carnage was huge. If we're living in an age roiling with tumultuous opportunity, those who have the right tools and concepts, and the discipline to employ them, will pull even farther ahead. Those who don't will fall farther behind. Many—despite the rich and robust opportunities—will get knocked out of the game entirely.[2]

Q: How did the 2008 financial meltdown affect your thinking for this study?

It served only to reinforce the relevance of the study question. Very few people predicted the 2008 financial crisis. The next Great Disruption will come, and the next one after that, and the next one after that, forever. We cannot know with certainty what they'll be or when they'll come, but we can know with certainty *that* they will come.

Q: Are you more or less optimistic and hopeful after conducting this study?

We're *much more* optimistic and hopeful. More than any of our prior research, this study shows that whether we prevail or fail, endure or die, depends more upon what we do than on what the world does to us. We take particular solace from the fact that every 10Xer made mistakes, even some very big mistakes, yet was able to self-correct, survive, and build greatness.

RESEARCH FOUNDATIONS

RESEARCH FOUNDATIONS: METHODOLOGY

We chose the matched-pair case method as an appropriate research approach. The essence of this method is to select pairs of comparable companies such that each company in the pair differs on a particular dimension (long-term performance in our case). To form each pair, we identified seven companies that had attained exceptional long-term performance in highly uncertain and chaotic industries (called "10X companies"). We then matched each 10X company with a comparison company that had a similar starting point (same industry, similar age and size), yet achieved only average performance. The resulting data set consists of 14 companies organized into seven contrast pairs. Using historical company chronologies that we created through an exhaustive data-collection effort, we then analyzed the variables that could explain the differences in long-term performance. Here are the steps we took.

1. **Identifying the Research Question and Unit of Analysis.** Our research question was, "*Why do some companies thrive in uncertainty, even chaos, and others do not?*" We classified an industry as highly uncertain and chaotic if it experienced a significant number of events that met these five criteria: (1) the events were out of the control of companies in the industry, (2) the events had an impact quite quickly (usually in much less than five years), (3) the events' impact could hurt companies in the industry, (4) some significant aspects of the events

were unpredictable (e.g., the timing, the form, the shape), and (5) the events actually happened (they weren't just predicted). The selected industries experienced tumultuous events that wreaked havoc, including deregulation, radical technology shifts, price wars, fuel shocks, regulatory and legal changes, consolidations, and industry recessions.

The unit of analysis in our study was not a company in perpetuity but a *company era*—the time from founding to June 2002, which was our observation period (our study overall covered the time period from approximately 1970 to 2002). Bounding the time frame was important, because we can't comment on what will happen to the companies *after* our study period. This era covered the company's start-up phase, its transition to a public company, its growth years, and its mature years as a large public enterprise.

2. **Selecting the Appropriate Research Method: the Matched-Pair Methodology.** We chose a methodology that would allow us to maximize the potential for discovering new insights that could be generalized across specific companies and industries: the *multiple-case research methodology* used in organizational behavior research. It is a comparative-case-method research design that is based on qualitative data collection and an inductive method of analysis. This approach relies on a small number of cases that can be studied in depth to identify patterns that form the basis of new findings.

In this method, researchers select cases that highlight differences in the variables of interest. The idea is that a contrast between the cases (companies) affords the best possibility to arrive at new findings. This approach follows a tradition in organizational behavior, finance, and medical research.[1] In their overview of this approach in the *Academy of Management Journal* in 2007, Kathy Eisenhardt and Melissa Graebner noted, "A particularly important theoretical sampling approach is 'polar types,' in which a researcher samples extreme (e.g., very high and very low performing) cases in order to more easily observe contrasting patterns in the data."[2] For example, in their study published in the *Academy of Management Journal* in 2010, Jeffrey Martin and

Kathy Eisenhardt selected high- and low-performing collaborative software teams, and analyzed factors that could explain the difference in performance.[3]

A key benefit of using the matched-pair method is that we avoid "sampling on success." If researchers study only successful companies, it becomes difficult to know whether the findings had anything to do with explaining that success. Perhaps losers followed the same management principles as the winners. To avoid this problem, we selected both successful *and* less successful companies, and studied the contrast.[4]

3. **Selecting the Study Population: Companies That Went Public in the United States.** We chose a study population such that the companies would feel the impact of uncertain and chaotic events around them, and not be insulated from those events because of sheer size or age. We selected companies from one population that fits this requirement—those that went public (had their initial public offering, or IPO) in the United States between 1971 and 1990. These were mostly young and/or small companies when they went public, and thus were fairly vulnerable to events in their environment.

4. **Identifying Exceptionally Performing Companies.** To compare companies across industries, we chose a performance measure, stock return, that applies equally across industries. (See *Research Foundations: 10X-Company Selections* for the precise measure.) This measure excludes different measures of performance that matter to other stakeholder groups, such as employees and communities. Nevertheless, it is perhaps the most important common metric for public companies. This measure also excludes other intermediate outcomes, such as innovation and sales growth. We view these measures, however, as possible *input* variables that might explain subsequent stock market performance.

Using stock-performance measures, we went through a systematic screening process and identified seven exceptionally performing companies (10X firms) in seven highly uncertain and chaotic industries drawn from our initial study population.

5. **Selecting Comparison Companies.** We used two overarching principles for selecting a comparison company for each of the seven 10X firms: (1) at the time when it became a public company, the comparison should have been similar to the 10X company (same industry, similar age, and similar size); and (2) it should have registered an *average* stock market performance (so as to create a contrast in performance between each 10X and comparison company). See *Research Foundations: Comparison-Company Selections* for details.

6. **Collecting Data: Historical Chronology.** We systematically went back in time and collected historical documentation for each company. For example, for Intel, we collected historical documentation for every year since its founding in 1968—company reports and press articles that appeared in 1968, 1969, 1970, 1971, and so on. We used a broad set of archival data sources to ensure that we obtained a comprehensive set of facts, views, and insights on the companies:

- All major articles published on each company *over our entire observation period* (from company founding date to 2002), from broad sources such as *Business Week*, the *Economist*, *Forbes*, *Fortune*, *Harvard Business Review*, the *New York Times*, the *Wall Street Journal*, and the *Wall Street Transcripts*; and from industry- or topic-specific sources
- Business-school case studies and industry analyses
- Books written about each company and/or its leaders
- Annual reports, proxy reports, and IPO prospectus for each company
- Major analyst reports on each company
- Business and industry reference materials, such as the *Biographical Dictionary of American Business Leaders* and the *International Directory of Company Histories*
- Materials obtained directly from each company (we wrote to them requesting information such as their corporate history,

speeches by senior executives, investor relations materials, and articles about the company)
- Company financial data: income and balance sheet data (from Compustat)

Consistent with qualitative-research methods, we examined a range of factors that could potentially explain the difference in outcomes between the 10X and comparison companies. This was a systematic effort to be open to possible novel explanations—the very purpose of using inductive case research. To this end, we collected information on a number of factors over time, which included the following:

- Leadership: key executives, CEO tenure and successions, leadership styles and behaviors
- Founding roots: founding team and circumstances
- Strategy: product and market strategies, business models, key mergers and acquisitions, strategic change
- Innovations: new products, services, technologies, practices
- Organizational structure, including significant reorganizations
- Organizational culture: values and norms
- Operating practices
- Human resource management: policies and practices related to hiring, firing, promotions, reward systems
- Use of technology, including information technology
- Company sales and profit trends, financial ratios
- Key industry events: downturns, booms, shocks, technology shifts, market shifts, regulatory changes, competitor moves, price wars, business-model changes, consolidations
- Major luck events (good and bad)
- Significant risk events
- Speed: time to spot threats and opportunities, time taken to make decisions, time to market (first mover or follower)

We then built a historical chronology by grouping all information on each company by year, starting with the earliest year and moving forward to 2002, the last year of our observations.

In building the historical chronology, we also searched for more than one source to verify each piece of information. This *triangulation of data* reduced the risk that our information was inaccurate, incomplete, or biased. For example, a book on PSA claimed that a Southwest Airlines team visited PSA in California in 1969 and was allowed to copy its operating manuals. We triangulated this information, and it was confirmed in another book by Southwest Airlines CEO Lamar Muse, who participated in the visit.[5]

In summary, our approach relied on gathering *high-quality data*. We followed rigorous academic principles for ensuring the integrity of the data, by gathering historical (not current) information dating back to the time of company founding, by including a breadth of data sources, by triangulating across sources, and by collecting data on a range of factors to avoid narrowing the inquiry up-front.

7. **Conducting Analysis.** *Within-pair analysis.* Once the historical chronology was constructed for a pair of companies, each one of us—Jim and Morten—separately read every single document, and wrote a detailed case report on each company and a pairwise case analysis. These pairwise documents averaged 76 pages (27,600 words) each, for a total of 1,064 pages (386,400 words) of case reports.

For each pair, we read each other's report. After a series of discussions, we generated a list of main *possible* explanations for the performance difference in a pair. A possible explanation had to meet the following criteria:

- A clear difference between the 10X and comparison companies, supported by compelling evidence
- An explanation for *why* this difference affected the outcome, known in academic research as a *causal mechanism* (the existence of a difference is not enough; there also needs to be a

plausible explanation for how the variable in question explains the difference in performance)

Cross-pair analysis. We looked for factors that were clearly present in most of the seven 10X companies and not in the comparison companies.

Concept generation. By drawing on the within-pair and cross-pair analyses, we identified the major concepts that seemed to explain the differences in outcomes. We made inferences from a set of individual factors and grouped them to develop more unifying concepts.

Financial analysis. We obtained data from Compustat and built detailed annual income, balance sheet, and cash flow statements from time of company founding (or earliest year with available data) to 2002, creating a spreadsheet with a total of 300 company-years of annual statements.

Event-history analysis. Deploying the event-history analysis method used by organizational scholars studying the evolution of companies, we analyzed the following events in a company's life: "20 Mile March" events, innovation events, "cannonball" events, risk events, time-sensitive events, and "SMaC recipe"–change events.[6] For each, we defined the term and coded for any occurrence by year, yielding an event history for each company (see the subsequent sections in *Research Foundations*).

8. **Limitations and Issues.** Every research method has its strengths and weaknesses. Ours is no exception. Here are the most common questions raised regarding it and our responses.

Isn't a study of 14 companies too small a sample?

No, because our aim wasn't to test existing hypotheses in a large sample of companies but rather to generate new findings. *The test of whether we had an adequate study set was whether we were confident that we had enough pairs to detect a pattern across them, such that if we added yet one more pair, we would likely not learn anything new.*[7] This is known in research methods as redundancy or theoretical saturation: at some point in qualitative case analysis (usually after 8 to 12 cases), the researcher reaches saturation, at which point no new knowledge is gained by simply adding more cases.[8] In our study, the final pair we added did not add further insight. One reason we reached saturation was that our deliberate, matched-pair design generated "polar types" that allowed us to discover differences more easily.

Is this sampling on success?

No, it isn't, as we explained earlier. We didn't select successful companies only. We selected contrasting pairs of companies in one industry, such that one company performed very well and another (the comparison) did not.

Can our findings be generalized?

Yes, they can, but with some qualifications:

- *Across many industries and companies.* We don't know whether our findings would hold across *all* companies. We are confident, however, that they are likely to hold across many companies and industries, because our findings are based on a diverse data set consisting of seven industries (and not just one or two industries). Also, because we studied U.S. companies only, one needs to be careful in extending the findings to other countries and cultures.
- *Across time.* Although we studied companies in the 1970–2002 era, we strongly believe that our findings are highly relevant for 2011 onward. The reason is that we purposefully selected highly uncertain and chaotic industries. To the extent that the world continues to be uncertain, what these industries experienced in terms of turmoil is likely to become the norm going

forward, making the insights derived from our research very relevant for the future.

Is there a potential bias in relying on recall of historical data?

This can be a potential problem, but our approach mitigates this issue. We have gone to great lengths to collect *historical* records, as opposed to current writings that interpret history by looking backward. For example, it is one thing to use an article on Intel from 2000 that looks back at its formative years in the 1970s; such an approach relies on historical *interpretation*, and this account might be colored by Intel's success in 2000. This approach is prone to problems known as *attribution errors*.[9] In contrast, we went back to *historical records* and collected information on events involving Intel *as they were happening* in the 1970s. At those times, the subsequent huge success of Intel had not occurred, so no one could make these attribution errors.

Can we claim causality?

Much of social science research, including most management research and ours, cannot claim causality if that term refers to deterministic causality ("a given change in *x* reliably produces a change in *y*"). Following a long tradition in organizational and strategic-management research, we instead seek to isolate explanations that *likely* led to performance differences between the companies. We've carefully chosen our language to reflect statements such as "There is likely an association between *x* and performance" and "An increase in *x* is likely to lead to an increase in *y*," which are probabilistic, not deterministic, statements.

Is there an issue of "reverse causality"?

Reverse causality occurs when the explanation is in the opposite direction from an initial hypothesis. For example, perhaps you initially thought that innovation had led to a company's success, when in fact, it was the company's success that led to better innovation (more successful firms have more money to invest in innovation). We largely avoided this potential problem because we relied on historical documents, through which we were able to identify *when* certain practices started, and thus we knew which factors came first.

Are there other companies that followed these principles and didn't attain the same level of success?

Since we haven't studied all companies in the United States, we can't verify whether this is the case. But the following makes this less of an issue:

- Our diverse data set (seven different industries) reduces the likelihood that our findings are just idiosyncratic to one or two companies and/or industries.
- As we said above, we are *not* claiming a deterministic causal chain—that if you adopt these principles, you will attain exceptional performance (guaranteed). We're stating that pursuing these principles *improves the probability of success.*
- Our 10X companies practiced *all* the principles articulated in this book; companies that practice one or a few would most likely not achieve exceptional performance.

Don't industry characteristics explain the outcomes?

We control for the impact of industry conditions by studying two companies in each industry (a matched pair). While both companies in a pair faced very similar industry situations, they nevertheless varied significantly in their practices and in their performances. Because industry factors are held constant for each pair, they alone can't explain these differences.

Weren't the 10X companies just lucky?

Critics of management research sometimes charge that the role of luck is often excluded in the analysis. Rather than ignoring the role of luck, we defined the construct, collected data on good- and bad-luck events, and examined the role these luck events played in explaining performance. We devote Chapter 7 to our findings about luck.

What if some of the companies don't perform well after our study period?

If this were to happen, it would not mean that what we found was invalid. Our claim is bounded; we studied company eras, not company performance in perpetuity. Performance may not last forever in these specific companies because of the following:

- The company may stop practicing the factors that led to its success.[10]
- Some redirection or new practices may be required after a very long run.
- Competitors may have caught up and copied a company's practices, rendering the original formula for success less potent.
- The stock market may have fully understood the company's success factors and thus accounted for them in the company's share price, making future extraordinary stock returns more difficult to achieve.

Any one of these explanations can cause a company's performance to erode. Just because great performance did not last, this does not invalidate the factors that helped create the performance in the first place.

RESEARCH FOUNDATIONS: 10X-COMPANY SELECTIONS

We used three overarching selection principles to identify our study set of exceptionally performing companies:

1. They achieved spectacular results; they were the clear winners in the stock market and their industry during our observation period.
2. They were in highly uncertain and chaotic industries.
3. They were vulnerable early on (being young and/or small companies that went public in 1971 or later).

We began with a data set drawn from the University of Chicago Center for Research in Security Prices (CRSP) database and went through the following 11 "filtering" steps to winnow it down.

SCREENING PROCESS TO IDENTIFY EXCEPTIONALLY PERFORMING COMPANIES

CUT 1. Begin with 20,400 companies that first appeared in CRSP 1971 or later. Eliminate those that first appeared after 1995 *15,852 companies left*
CUT 2. Companies still in existence after June 2002 *3,646 companies left*

CUT 3. TSR performance at least 3x by 2002*   *368 companies left*
CUT 4. Real U.S. company IPO 1971–90   *187 companies left*
CUT 5. Exclude small companies as of 2001   *124 companies left*
CUT 6. TSR performance at least 4x 15 years after IPO*   *50 companies left*
CUT 7. Eliminate inconsistent performance patterns   *25 companies left*
CUT 8. Uncertain and chaotic industries only   *12 companies left*
CUT 9. Red flag test (concerns)   *9 companies left*
CUT 10. Exclude too large and old at IPO   *8 companies left*
CUT 11. Outperform industry   *7 companies left*

* Company cumulative return ratio to the market (see "Key Definitions")

Cut 1: Select companies first appearing in CRSP 1971–95. We reasoned that a first data entry in CRSP was a good proxy for when a company went public (see Cut 4).[11]

Cut 2: Keep companies in existence after June 2002. We wanted to include only companies that were ongoing, independent concerns at the end of the observation period in 2002.

Cut 3: Meet initial stock performance threshold. We eliminated all companies where the company's monthly cumulative return ratio to the market fell below 3.0, based on the time from a company's first CRSP end-of-month date to June 28, 2002 (see "Key Definitions").[12]

KEY DEFINITIONS

▶ ***Monthly Total Return:*** The total return to shareholders in a given month, including dividends reinvested, for an individual security (also called total shareholder return, TSR).

▶ ***Cumulative Stock Return:*** The compounded value of $Y invested in an individual security between times t1 and t2, using the formula $Y x (1 + Monthly Total Return @ m1) x (1 + Monthly Total Return @ m2) x . . . (1 + Monthly Total Return @ t2); where m1 = end of the first month following t1, m2 = end of the second month following t1, and so forth.

▶ ***General Stock Market*** (also called *general market* or just the *market*): NYSE/AMEX/NASDAQ value-weighted return, which consists of the combined market value of all companies traded on these exchanges (including dividends reinvested), weighted by the capitalization of the company divided by the capitalization of the market.

▶ ***Cumulative Return Ratio to the Market:*** At the end of any given time period, this ratio is calculated as the cumulative return of $Y invested in the company divided by the cumulative return of $Y invested in the general stock market, such that $Y is invested in both the company and the market on the same date.

Note: We used the same formulas for Cut 6, replacing monthly- with daily-return data.

Cut 4: Verify were real U.S. companies with IPOs 1971–90. We performed due diligence on every remaining company to verify when the company went public and that it was indeed a real company. We eliminated non-traditional IPOs such as spinouts, reverse mergers, mergers, reverse LBOs, REITs, and limited partnerships. We also eliminated foreign companies.

*Cut 5: **Eliminate companies with less than $500 million in revenue as of 2001.*** While we wanted to analyze young and/or small companies in their early years, we also needed to analyze companies that had grown into large companies by the end of the observation period.

*Cut 6: **Meet stock-performance threshold from IPO date to 15 years afterward.*** We used a more precise and stringent stock-performance criterion based on *daily-return* data for the period from a company's IPO date to 15 years afterward. We eliminated all companies where a company's cumulative return ratio to the market for this period fell below 4.0.

*Cut 7: **Eliminate companies with inconsistent stock-performance patterns.*** The purpose of this cut was to eliminate companies that showed inconsistent stock-performance patterns (e.g., erratic, up and down).

*Cut 8: **Select companies in highly uncertain and chaotic industries.*** We classified an industry as highly uncertain and chaotic if it experienced a significant number of events that met the following five criteria:

1. The events were out of the control of companies in the industry; they couldn't prevent them from happening.
2. The events had an impact quite quickly. For our purposes, "quickly" meant less than five years. (Usually, they happened much faster than that.)
3. The events' impact could hurt companies in the industry. They might not have hurt every single company (including the company under consideration), but they had the potential to hurt them.
4. Some significant aspects of the events were *unpredictable*. The events themselves might not have been entirely unpre-

dictable, but some important elements of the events were unpredictable—the timing, the form, the shape, the impact, etc. (For instance, deregulation in the airline industry was predictable, but the exact form that deregulation took and how it affected the industry shakeout was not entirely predictable.)

5. The events actually happened; they weren't just predicted.

We systematically collected information on the industries and created coding documents for each industry. Using these analyses, we categorized industries as "stable," "moderately uncertain," and "highly uncertain and chaotic." We selected companies whose industries fell in the latter category.

Cut 9: Red Flag test. We conducted a "red flag" analysis to identify whether the company had experienced a significant restatement of earnings during the observation period and/or was fundamentally weak at the time of final selection. We excluded cases of concern.

Cut 10: Young or small at IPO. Because we wanted only companies that were either young or small at the time of IPO, we eliminated those that were both old and large at that time.

Cut 11: Outperform industry index. The purpose of this test was to ensure that the companies did not simply perform well because their industry performed well. We created industry stock-performance indices and excluded a company if its cumulative stock return did not outperform that of its industry by 3x from the date of the company's IPO to 15 years afterward.

RESEARCH FOUNDATIONS: COMPARISON-COMPANY SELECTIONS

Using historical documents, we conducted a systematic search to identify industry peers, scored each of them, and selected the best match. We scored comparison candidates on the following six criteria. Criteria 1 to 4 ensure that the comparison had a similar starting point as the 10X company; Criterion 5 creates the performance gap; and Criterion 6 is a face-validity check. The final comparison choices rate as excellent or very good based on our criteria, with one exception (Kirschner, which was acceptable).

1. **Business fit (early years).** The 10X company and the comparison candidate were in similar businesses at the time when the 10X company went public (practically speaking, we used the year of first available stock returns in CRSP, hereafter called the "match year").[13]

2. **Age fit.** The comparison candidate was founded around the same time as the 10X company.

3. **Size fit (early years).** The two companies were of similar size at the time when the 10X company went public.

4. Conservative test (early years). At the time when the 10X company went public, the comparison candidate was *more successful* than the 10X company (it is a more stringent screen to have an initially strong comparison).

5. Performance gap. The comparison candidate's cumulative return ratio to the market (referred to as "ratio" below) was close to or below 1.0 during our selection period (i.e., the comparison candidate's shareholder return was no better than the general stock market during this time).[14]

6. Face validity (in 2002). The comparison candidate "makes sense" when looking at the two companies at the end of our observation period; they continued to be in similar businesses.

SUMMARY REMARKS FOR EACH MATCHED PAIR

Amgen. Number of biotech firms considered: 12. Best match: *Genentech.* Match year: 1983. Excellent match on conservative test, face validity, business fit, and performance gap (ratio 1983–2002 = 0.92). Weaker match on age fit and size fit. Comment: Genentech was an early leader in the biotech industry (founded in 1976), while Amgen was one of several new biotech companies formed in 1980. Runners-up: Chiron, Genzyme.

Biomet. Number of orthopedic medical-device makers considered: 10. Best match: *Kirschner.* Match year: 1986. Very good match on business fit, size fit, and performance gap (ratio 1986–94 = 0.76). Weaker match on conservative test, face validity, and age fit. Comment: Kirschner and Biomet both focused on the orthopedic-implant and reconstructive-device markets. Runners-up: Advanced Neuromodulation Systems, Intermedics.

Intel. Number of integrated-circuit firms considered: 16. Best match: *Advanced Micro Devices (AMD).* Match year: 1973. Excellent match on business fit, age fit, face validity, and performance gap (ratio 1973–2002 = 1.05). Weaker match on conservative test and size fit. Comment: Both Intel and AMD were started by men who left Fairchild Semiconductor in the late 1960s and focused on memory chips. Runners-up: Texas Instruments, National Semiconductor.

Microsoft. Number of computer firms considered: 10. Best match: *Apple.* Match year: 1986. Excellent match on age fit, face validity, and performance gap (ratio 1986–2002 = 0.51). Weaker match on business fit, conservative test, and size fit. Comment: During our key observation years (late 1970s to mid-1990s), Microsoft and Apple offered two alternative personal computer platforms and were competitors. Runners-up: Lotus, Novell.

Progressive. Number of insurers considered: 16. Best match: *Safeco.* Match year: 1973. Excellent match on business fit, conservative test, and performance gap (ratio 1973–2002 = 0.95). Weaker match on face validity, size fit, and age fit. Comment: Like Progressive, Safeco was long a premier auto insurer with underwriting discipline. Runners-up: GEICO, Employers Casualty.

Southwest Airlines. Number of airlines considered: 25. Best match: *Pacific Southwest Airlines (PSA).* Match year: 1973. Excellent match on business fit, conservative test, face validity, and performance gap (ratio 1973–87 = 0.99). Weaker match on size fit and age fit. Comment: Southwest Airlines copied its business model directly from PSA. Runners-up: Braniff, Continental/Texas.

Stryker. Number of surgical-device firms considered: 15. Best match: *United States Surgical Corporation (USSC).* Match year: 1979. Excellent match on business fit, conservative test, face validity, and performance gap (ratio 1979–98 = 1.16). Weaker match on age fit and size fit. Comment: From the 1970s onward, both USSC and Stryker focused on surgical instruments and equipment. Runners-up: Birtcher, American Hospital Supply.

RESEARCH FOUNDATIONS: 20 MILE MARCH ANALYSIS

As discussed in Chapter 3, we coded for and analyzed the companies' 20 Mile March behaviors; i.e., whether they had markers that delineated lower bounds for performance and self-imposed constraints to hold back during good times. We catalogued whether the companies articulated and achieved such practices, and we also analyzed the effects of adhering to the 20 Mile March principle on company outcomes in 52 industry-downturn events.

Finding 1. The 10X companies practiced the 20 Mile March principle to a much greater extent than the comparison companies (strong evidence). There was strong support for this in six out of seven pairs and good support in one pair (Amgen and Genentech). Two comparison companies (PSA and Safeco) started out adhering to the 20 Mile March approach but then neglected it over time. Two comparison companies, Genentech and Apple, adopted a 20 Mile March approach later on. The other comparison companies (USSC, Kirschner, and AMD) showed little evidence of having a 20 Mile March approach (see the "20 Mile March Contrasts through 2002" table in Chapter 3).

Finding 2. Companies that practiced the 20 Mile March principle at a given time performed much better in subsequent industry downturns than those that didn't (strong evidence). There was strong support for this finding in all seven pairs. Several comparison companies that did not adhere to the 20 Mile March practice fared poorly during industry downturns.

As the following table reveals, there was a very large benefit from 20 Mile Marching *before* a difficult time in the industry. Practicing 20 Mile Marching was far more often associated with subsequent good outcomes (29 events) than poor outcomes (0) in difficult times. *Not* taking a 20 Mile March approach was far more often associated with poor outcomes (20) than good outcomes (3).

As shown in the table, the comparison companies also benefited from 20 Mile Marching during the few times (4) they practiced it. Also, the few times (2) the 10X companies failed to practice the 20 Mile March, the outcomes were negative. *The 10X companies did much better in difficult industry times because they adhered to the 20 Mile March approach beforehand, while the comparison companies suffered poor performance in difficult industry times because they most often did not adhere to this practice.*

20 MILE MARCH PRACTICE AND OUTCOME DURING INDUSTRY DOWNTURNS

Type of Combination (20 Mile March + Outcome)	Number of Events (%)		
	10X Companies	*Comparison Companies*	*Total*
Industry Downturn Events	27	25	52
20 Mile March Practice	25 (100%)	4 (100%)	29 (100%)
20 Mile March + Good Outcome	25 (100%)	4 (100%)	29 (100%)
20 Mile March + Poor Outcome	0 (0%)	0 (0%)	0 (0%)
No 20 Mile March Practice	2 (100%)	21 (100%)	23 (100%)
No 20 Mile March + Good Outcome	0 (0%)	3 (15%)	3 (13%)
No 20 Mile March + Poor Outcome	2 (100%)	18 (85%)	20 (87%)

N=52 industry downturns

Note: Comparable numbers of years coded for 10X and comparison companies.

RESEARCH FOUNDATIONS: INNOVATION ANALYSIS

As discussed in Chapter 4, we performed an analysis of 290 innovation events to determine the types and degree of innovation among the 10X and the comparison companies.

The term "innovation" is a multifaceted construct. First, innovation refers to different dimensions, including product, operational, and business-model innovations. What constitutes critical innovation depends on the industry.

Second, much has been written about degrees of innovativeness.[15] A radical or revolutionary innovation has a very large performance or feature improvement compared to existing offerings, while an incremental or evolutionary innovation has a small performance or feature improvement. We coded innovations according to whether they were incremental, medium, or major. By an *innovative* company, we meant one that had several *major* and *medium* innovations.

Third, several reference points can be used—innovative compared to what? One reference point is relative to what the company had offered previously (an internal reference point). Another reference point is relative to what existed in the marketplace at that time (an external reference point). We adopted the latter viewpoint.

Fourth, it's possible to have a very innovative product that isn't a commercial success. It is important not to confuse innovation with the financial outcome in the market.

We began by identifying the most important areas of innovation in each industry. We also judged the *innovation threshold* in each industry—to what extent the nature of the industry required a company to be innovative just to be a player. While some industries have high thresholds (e.g., biotech), others have low thresholds (e.g., airlines).

We coded innovation events by analyzing historical company and press documents to identify announcements of innovations.[16] To code for the degree of innovativeness, we created the following categories:

- *Major innovation.* The innovation clearly offered a high degree of performance or feature improvement compared to existing products/services in the marketplace. Often called "pioneering," "revolutionary," or "breakthrough."
- *Medium innovation.* The innovation offered a solid degree of performance or feature improvement.
- *Incremental innovation.* The innovation offered some performance or feature enhancement, but it clearly didn't signify major progress.

Finding 1. The companies in our study created a number of innovations during our observation period (good evidence). Overall, we counted 290 innovation events across the companies: 31 major, 45 medium, and 214 incremental ones (see the following table). Twelve companies clearly developed a number of innovations during the study period. Two, Safeco and Kirschner, did not.

Finding 2. There appears to be an innovation "threshold" effect: companies innovated more in industries in which innovation played a significant role (good evidence). Companies in high-threshold industries (biotech, semiconductors, personal computers) created on average 7.5 major/medium innovations during our observation period, while this number was 5.0 for medium-threshold industries (medical devices) and 2.8 for low-threshold ones (airlines, auto insurance).

Finding 3. The 10X companies were *not* more innovative than the comparison companies (strong evidence). The following table shows no clear pattern across the pairs. Three 10X companies were clearly more innovative by having a higher number of major/medium innovations (Intel more than AMD; Progressive more than Safeco; Biomet more than Kirschner). In the other four pairs, it appears to be the other way around; the comparison companies were more innovative than the 10X companies (PSA over Southwest Airlines; Genentech over Amgen; USSC over Stryker; Apple over Microsoft).

(In biotech, patents can be used to indicate innovativeness. According to data provided by the U.S. Patent Office, Genentech had many more patents issued [772] than Amgen [323] from founding to 2002.[17] In addition, according to patent data provided by INSEAD professor and patent-data expert Jasjit Singh, Genentech's patents were also more cited by other patents, a measure of degree of innovation: average citations per patent were 7.09 for Genentech versus 4.23 for Amgen.[18] Thus, Genentech was more innovative based on the patent measure, confirming our innovation count.)

Finding 4. The 10X companies pursued more incremental innovations than the comparison companies (some evidence). In five out of seven pairs, the 10X companies had higher incremental innovation counts than their comparisons (see the last column in the following table). This tendency ties in with the notion of the 20 Mile March: companies that adhere to a practice of taking "small steps of progress every day" are likely to emphasize small but frequent innovations.

SUMMARY OF INNOVATION-EVENT COUNTS

		10X Company			Comparison Company			10X More Innovative?*	10X More Incremental?
Matched Pairs	*Industry Innovation Threshold*	*Number of Major*	*Number of Medium*	*Number of Incremental*	*Number of Major*	*Number of Medium*	*Number of Incremental*	*10X vs. Comparisons*	*10X vs. Comparisons*
Intel and AMD	High	4	6	15	1	4	11	Yes: 10 vs. 5	Yes: 15 vs. 11
Amgen and Genentech	High	2	2	8	6	2	4	No: 4 vs. 8	Yes: 8 vs. 4
Microsoft and Apple	High	2	6	23	6	4	14	No: 8 vs. 10	Yes: 23 vs. 14
Biomet and Kirschner†	Medium	2	3	4	0	0	2	Yes: 5 vs. 0	Yes: 4 vs. 2
Stryker and USSC‡	Medium	1	6	77	3	5	41	No: 7 vs. 8	Yes: 77 vs. 41
Southwest Airlines and PSA	Low	1	2	3	2	3	7	No: 3 vs. 5	No: 3 vs. 7
Progressive and Safeco	Low	1	2	2	0	0	3	Yes: 3 vs. 0	No: 2 vs. 3
Median		2	3	8	2	3	7		
Total		13	27	132	18	18	82	3 Yes, 4 No	5 Yes, 2 No

N=290 innovation events. Note: Comparable numbers of years coded for 10X and comparison companies.

* More Innovative = having the largest number of significant innovations, which are defined as the sum of the number of major innovations and the number of medium innovations.

† Incomplete information.

‡ Both companies until 1997 only.

RESEARCH FOUNDATIONS: BULLETS-THEN-CANNONBALLS ANALYSIS

The discussion in Chapter 4 is based on our analysis of the prevalence of a bullets-then-cannonballs approach and the outcome of 62 cannonball events for the 10X and comparison companies. We conducted an event-history analysis by identifying, counting, and analyzing bullets and cannonballs.[19]

Finding 1. The 10X companies pursued more of a bullet approach than the comparison companies (good evidence). The 10X companies practiced the bullet approach more than their comparison companies in five out of the seven pairs. In two pairs, the companies practiced it at the same level (Southwest Airlines and PSA; Amgen and Genentech).

Finding 2. The 10X companies did *not* fire more cannonballs than the comparison companies (strong evidence). As shown in the following table (Column 1), the comparison companies fired more cannonballs in five pairs, while the opposite was true in two pairs (Intel more than AMD; Progressive more than Safeco).

COUNT OF CANNONBALLS

Column #	1	2	3	4
Company	Number of Cannonballs	Number of Calibrated	Number of Uncalibrated	Calibrated* (%)
Southwest Airlines	5	4	1	80%
PSA	8	0	8	0%
Intel	7	5	2	71%
AMD	6	3	3	50%
Biomet	1	0	1	0%
Kirschner	3	0	3	0%
Progressive	4	3	1	75%
Safeco	3	0	3	0%
Amgen	3	2	1	67%
Genentech	4	2	2	50%
Stryker	2	1	1	50%
USSC	5	1	4	20%
Microsoft	4	3	1	75%
Apple	7	2	5	29%
Average for 10X Companies	3.7 (Total = 26)	2.6 (Total = 18)	1.1 (Total = 8)	69%
Average for Comparison Companies	5.1 (Total = 36)	1.1 (Total = 8)	4.0 (Total = 28)	22%

N=62 cannonball events

Note: Comparable numbers of years coded for 10X and comparison companies.

* (Number in Column 2)/(Number in Column 1)*100.

Finding 3. The 10X companies had a higher proportion of cannonballs that were calibrated than the comparison companies (strong evidence). As shown in Column 4 in the table above, when using cannonballs, the 10X companies deployed calibrated ones 69 percent of the time whereas the comparison companies did so only 22 percent of the time (remember that calibration means that the company had conducted an empirical trial beforehand to validate the initiative).

Finding 4. Calibrated cannonballs yielded more positive outcomes than uncalibrated ones (strong evidence). Of all the calibrated cannonballs launched, 88 percent resulted in a good outcome (see the following table). In stark contrast, only 23 percent of the *uncalibrated* ones yielded a good outcome. (Calibration is an activity that takes place *before* betting big; there's no guarantee, however, that calibration leads to success.)

CANNONBALL CALIBRATION AND OUTCOME (ALL COMPANIES)

Type of Outcome	Number of Calibrated Cannonballs (%)	Number of Uncalibrated Cannonballs (%)	Total Number of Cannonballs
Number of Positive Outcomes	23 (88%)	7 (23%)	30
Number of Negative Outcomes	3 (12%)	23 (77%)	26
Total	26 (100%)	30 (100%)	56

N=56 (excludes six cannonballs with unclear outcome)

Note: Comparable numbers of years coded for 10X and comparison companies.

Finding 5. The 10X companies had more success with their cannonballs than the comparison companies, principally because they launched more calibrated ones (strong evidence). As the next table reveals, of the 26 cannonballs that the 10X companies launched, 18 were calibrated, and 17 of those were successful. In contrast, the comparisons launched only 8 calibrated cannonballs (out of 36), and 6 of those were successful. *The comparison companies had a low chance of success with their cannonballs because so many of them were uncalibrated.*

CANNONBALL CALIBRATION AND OUTCOME

Type of Cannonball	Outcome	10X Companies	Comparison Companies	Total Outcomes
Calibrated	Number of Good Outcomes (%)	17 (94%)	6 (75%)	23 (88%)
	Number of Poor Outcomes (%)	1 (6%)	2 (25%)	3 (12%)
	Number of Calibrated (%)	18 (100%)	8 (100%)	26 (100%)

(continued on next page)

Type of Cannonball	Outcome	10X Companies	Comparison Companies	Total Outcomes
Uncalibrated	Number of Good Outcomes (%)	3 (37%)	4 (18%)	7 (23%)
	Number of Poor Outcomes (%)	5 (63%)	18 (82%)	23 (77%)
	Number of Uncalibrated (%)	8 (100%)	22 (100%)	30 (100%)

N=56 (excludes six observations with unclear outcome)

Note: Comparable numbers of years coded for 10X and comparison companies.

RESEARCH FOUNDATIONS: CASH AND BALANCE-SHEET-RISK ANALYSIS

As discussed in Chapter 5, we conducted an analysis of 300 company-years of financial statements to determine the extent to which the 10X and comparison companies built cash reserves and used debt.

Using Compustat data, we analyzed the following financial ratios for each matched pair on an annual basis and determined how frequently each 10X company had a better ratio than its comparison company. For cash, a higher ratio was considered better, and for debt, a lower ratio was considered better.

- Current ratio = (current assets)/(current liabilities)
- Cash to total assets = (cash and cash equivalents)/(total assets)
- Cash to current liabilities = (cash and cash equivalents)/(current liabilities)
- Total debt to equity = (long-term debt + current liabilities)/(stockholders' equity)
- Long-term debt to equity = (long-term debt)/(stockholders' equity)
- Short-term debt to equity = (current liabilities)/(stockholders' equity)

Finding 1. The 10X companies overall had more conservative balance sheets during the observation period than the comparison companies (strong evidence). As the following table shows, the 10X companies overall had more years with better cash and debt ratios than the comparison companies during the observation period ("All Years" column). They took less risk according to these measures.

Finding 2. The 10X companies overall had more conservative balance sheets in their *first five years* as public companies than the comparison companies (strong evidence). Finding 1 could simply have been because the 10X companies performed better (and thus could afford to have stronger balance sheets). But as the following table reveals, the 10X companies overall had better financial ratios than their comparisons in their first 5 years as public companies (as well as in the first 10 years). They took less risk early on according to these measures.

Finding 3. The 10X companies overall had more conservative balance sheets in their *first year* as public companies than the comparison companies (fairly good evidence). If we look at their *first year* as public companies ("IPO Year" column in the following table), the 10X companies had better cash ratios than the comparisons, and they performed better regarding two of the debt ratios, with the long-term debt ratio being equal across groups (three comparisons—PSA, Genentech, Apple—had lower debt than their corresponding 10X companies during their respective IPO years).

COMPARISON OF FINANCIAL RATIOS (ALL COMPANIES)

Area	Ratio	% of Time the 10X Companies Had Better Ratio Than the Comparison Companies				Which Did Better?
		All Years*	5 Years†	10 Years‡	IPO Year§	
Cash	Current ratio	72%	83%	72%	83%	10X
	Cash to total assets	80%	83%	80%	67%	10X
	Cash to current liabilities	80%	90%	80%	83%	10X
Debt	Total debt to equity	64%	80%	80%	67%	10X
	Long-term debt to equity	61%	61%	67%	50%	Mixed
	Short-term debt to equity	64%	87%	78%	100%	10X

* All Years = from first year during which both 10X and comparison companies became public and financial data were available to 2002 (comparable number of years for 10X and comparison companies).

† 5 Years = from IPO year to 5 years afterward.

‡ 10 Years = from IPO year to 10 years afterward.

§ IPO Year = first fiscal year in which the companies became public.

RESEARCH FOUNDATIONS: RISK-CATEGORY ANALYSIS

The discussion of risk categories in Chapter 5 is based on the following analysis of 114 decision events.

We analyzed the following types of risks:

- *Death Line risk:* This could kill or severely damage the enterprise.
- *Asymmetric risk:* The potential downside is much bigger than the potential upside.
- *Uncontrollable risk:* This exposes the enterprise to forces and events that it has little ability to manage or control.

Finding 1. The 10X companies overall made fewer decisions involving Death Line risk than the comparison companies (strong evidence). The comparison companies made an average of 2.9 decisions involving Death Line risk (36 percent of decisions, or nearly 4 out of 10), compared to only 0.9 such decisions (10 percent, or 1 out of 10) made by the 10X companies (see the following table).

Finding 2. The 10X companies overall made fewer decisions involving asymmetric risk than the comparison companies (strong evidence). While 36 percent of comparison-company decisions involved this type of risk, only 15 percent of 10X-company decisions involved asymmetric risk.

Finding 3. The 10X companies overall made fewer decisions involving uncontrollable risk than the comparison companies (strong evidence). The percentage of decisions involving uncontrollable risk was substantially lower for the 10X companies (42 percent) than for the comparisons (73 percent).

Finding 4. The 10X companies overall made less risky decisions (strong evidence). As the following table shows, 56 percent of decisions made by the 10X companies were low risk, compared with only 22 percent for the comparison companies (a low-risk decision doesn't involve *any* of the three types of risks outlined above). In contrast, 43 percent of decisions made by the comparison companies were high risk, compared with only 22 percent for the 10X companies.

TYPE AND EXTENT OF RISK INVOLVED IN IMPORTANT DECISIONS

Type of Decisions Made	10X Companies	Comparison Companies	10X or Comparison Took More Risk?
Average Number of Decisions Analyzed Per Company	8.4	7.9	
Decisions Involving Death Line Risk, % (Average Number)	10% (0.9)	36% (2.9)	Comparison
Decisions Involving Asymmetric Risk, % (Average Number)	15% (1.3)	36% (2.9)	Comparison
Decisions Involving Uncontrollable Risk, % (Average Number)	42% (3.6)	73% (5.7)	Comparison
Decisions Classified as Low Risk* (%)	56%	22%	Comparison
Decisions Classified as Medium Risk† (%)	22%	35%	
Decisions Classified as High Risk‡ (%)	22%	43%	
	100%	100%	

N=114 decisions

Note: Comparable numbers of years coded for 10X and comparison companies. Death Line Risk, Asymmetric Risk, and Uncontrollable Risk are not mutually exclusive categories (percentage refers to proportion of all decisions analyzed). Low Risk, Medium Risk, and High Risk are mutually exclusive categories.

* Low Risk = no Death Line Risk, no Asymmetric Risk, no Uncontrollable Risk.

† Medium Risk = no Death Line Risk, but one of either Uncontrollable Risk or Asymmetric Risk.

‡ High Risk = Death Line Risk and/or both Asymmetric Risk and Uncontrollable Risk.

Finding 5. The 10X companies had a higher success rate in all risk categories (good evidence). As the next two tables reveal, for low-risk decisions, the 10X companies succeeded 85 percent of the time (versus 64 percent of the time for the comparison companies). For medium-risk decisions, the 10X companies had a 70 percent success rate (versus 50 percent for the comparison companies). For high-risk decisions, the 10X companies succeeded 45 percent of the time (versus 5 percent of the time for the comparison companies). The high-risk contrast is striking. The main reason for this is that these decisions involved major bets—cannonballs. As we saw in the Bullets-Then-Cannonballs analysis, the 10X companies spent more time experientially validating those bets (by firing bullets) before going ahead, increasing the chances of success.

DECISION RISK AND OUTCOME
(10X COMPANIES ONLY)

	Risk Taken		
Outcome	**Low (%)**	**Medium (%)**	**High (%)**
Poor	0%	15%	55%
OK	15%	15%	0%
Success	85%	70%	45%
	100%	100%	100%

N=59 decisions

DECISION RISK AND OUTCOME
(COMPARISON COMPANIES ONLY)

	Risk Taken		
Outcome	**Low (%)**	**Medium (%)**	**High (%)**
Poor	18%	28%	75%
OK	18%	22%	20%
Success	64%	50%	5%
	100%	100%	100%

N=55 decisions

RESEARCH FOUNDATIONS: SPEED ANALYSIS

As discussed in Chapter 5, we analyzed 115 time-sensitive moments to determine the 10X and comparison companies' speed of recognition, deliberation, decision, and action.

We defined *unequal moments* as events where there are signs that conditions have changed and the risk profile is changing with time.

CLASSIFICATION OF UNEQUAL MOMENTS (ALL COMPANIES)

Facet	Characteristic (%)	
Pace of Events	Slow-Moving*: 30%	Fast-Moving: 70%
Nature of Moment†	Threat: 79%	Opportunity: 21%
Clarity of Response‡	Clear: 42%	Unclear: 58%
Outcome§	Good: 68%	Poor: 32%

N=115 moments

* Slow-moving = moment unfolded over a long period (one to three years typically).

† Fourteen moments were not classified.

‡ Clarity = pretty obvious what the company's response should have been (no need to deliberate for long).

§ There were 13 moments with unclear/OK outcomes.

Finding 1. Early recognition of an unequal moment was associated with a good outcome (strong evidence). As shown in the following table, in cases with good outcomes, the moments were recognized early 71 percent of the time (versus only 28 percent for cases with poor outcomes).

TIME OF RECOGNITION OF UNEQUAL MOMENT AND OUTCOME (ALL COMPANIES)

Time of Recognition	Good Outcome (%)	Poor Outcome (%)
Early*	71%	28%
Late	13%	66%

N=101 moments (excludes observations with insufficient information)

Note: Medium Time of Recognition category left out (100% = Early + Medium + Late).

* Early = the company recognized the first signals that the unequal moment was forming.

Finding 2. The benefit of fast decision making depended on the pace of events (fairly good evidence). Overall, fast decision making was associated with good outcomes (see the following table). This was even more pronounced in the case of fast-moving events. However, when events were moving slowly, 61 percent of cases with good outcomes involved a slow/medium decision-making speed. In other words, in good-outcome cases, *decisions were not always made quickly*; a fair number were made at a slow pace, when the events allowed. This suggests a "fast when you must, slow when you can" approach.

DECISION SPEED AND OUTCOME (ALL COMPANIES)

Speed of Events	Decision Speed	Good Outcome
All Observations (N = 98)	Slow/Medium (%)	35%
	Fast* (%)	65%
Fast-Moving Events (N = 69)	Slow/Medium (%)	25%
	Fast* (%)	75%
Slow-Moving Events (N = 29)	Slow/Medium (%)	61%
	Fast* (%)	39%

N=98 (excludes observations with insufficient information)

* Fast = decision was made quickly once the moment was recognized.

Finding 3. Deliberate decision making was associated with good outcomes (strong evidence). By deliberate, we mean that there was evidence that the leaders took a step back, *zoomed out*, and considered at a deeper level why things were happening. In contrast, when using the term "reactive," we mean decision making that lacked rigorous deliberation, situations in which the leaders either followed convention or made impulsive decisions. As the following table shows, 63 percent of the good-outcome cases were associated with a deliberate approach, while a full 97 percent of the bad-outcome cases were associated with a reactive approach.

DELIBERATE VERSUS REACTIVE APPROACHES AND OUTCOME (ALL COMPANIES)

Type of Decisions	Good Outcome (%)	Poor Outcome (%)
Deliberate	63%	3%
Reactive	37%	97%

N=100 (excludes observations with insufficient information)

Finding 4. The benefit of fast execution depended on the pace of events (good evidence). Overall, fast execution was associated with good outcomes (see the following table). This was even more pronounced with fast-moving events, with 81 percent of the good-outcome cases associated with fast execution. With slow-moving events, the picture is mixed: both fast and slow/medium executions were associated with good outcomes.

EXECUTION SPEED AND OUTCOME (ALL COMPANIES)

	Execution Speed	Good Outcome (%)
All Observations (N = 65)	Slow/Medium (%)	27%
	Fast* (%)	73%
Fast-Moving Events (N = 46)	Slow/Medium (%)	19%
	Fast* (%)	81%
Slow-Moving Events (N = 19)	Slow/Medium (%)	50%
	Fast* (%)	50%

N=65 (There are fewer observations here because many times companies did not change anything, so no execution followed.)

* Fast = leaders implemented quickly once decision was made.

Finding 5. The 10X companies adhered to Findings 1 to 4 more than the comparison companies (strong evidence).

- *Time of recognition.* The 10X companies recognized the emergence of an unequal moment early in a greater proportion of cases (68 percent) than the comparisons (42 percent).
- *Decision speed.* Overall, the 10X companies made decisions quickly in a greater proportion of cases (57 percent) than the comparisons (45 percent). However, they were also much better at *moderating* decision speed: their proportion of fast decision making increased to 71 percent for quick-paced events (versus 52 percent for the comparisons). And this dropped to 25 percent for slow-moving events (versus 31 percent for the comparisons).
- *Deliberate versus reactive.* The 10X companies were deliberate in a higher proportion of the decisions (68 percent) than the comparisons (14 percent).
- *Execution speed.* Overall, the 10X companies *did not* execute quickly in a substantially higher proportion of cases (66 percent) than the comparisons (63 percent). However, they were much better at *moderating* execution speed: the proportion of fast-execution decisions increased to 76 percent for quick-paced events (versus 62 percent for the comparisons). And this dropped to 40 percent for slow-moving events (versus 67 percent for the comparisons).

As a result, the 10X companies had a greater proportion of unequal moments with good outcomes (89 percent) than the comparison companies (40 percent).

DECISION-RELATED BEHAVIORS ADHERED TO BY THE 10X AND COMPARISON COMPANIES DURING UNEQUAL MOMENTS

Aspect of Unequal Moment		10X Companies (N = 57)	Comparison Companies (N = 45)
Time of Recognition	Recognized Early (%)	68%	42%
Decision Speed	Fast Decision Making (%)	57%	45%
	Fast Decision Making for Fast-Moving Events (%)	71%	52%
	Fast Decision Making for Slow-Moving Events (%)	25%	31%
Deliberate vs. Reactive	Deliberate Decision Making (%)	68%	14%
Execution Speed	Fast Execution (%)	66%	63%
	Fast Execution for Fast-Moving Events (%)	76%	62%
	Fast Execution for Slow-Moving Events (%)	40%	67%

N=102 moments (excludes observations with insufficient information)

Note: Comparable numbers of years coded for 10X and comparison companies. 100% = all observations in that category for the 10X (comparison) companies.

RESEARCH FOUNDATIONS: SMaC-RECIPE ANALYSIS

As discussed in Chapter 6, we analyzed each company to identify the extent to which it had a SMaC recipe, and if so, compiled its elements. We recorded 117 SMaC elements across all the companies; when they originated; whether they changed; and if so, when.

Finding 1. The 10X companies had clearly understood SMaC recipes (strong evidence). All seven 10X companies had a fully formed SMaC recipe as a young and/or small company.

Finding 2. The comparison companies had clearly understood SMaC recipes (fairly good evidence). Five comparison companies (PSA, Safeco, Apple, Genentech, and USSC) had clearly established SMaC recipes as young and/or small companies, while one company (AMD) had a vague recipe and one (Kirschner) never had one.

Finding 3. The 10X companies rarely changed the elements of their SMaC recipes (strong evidence). As the table below reveals, the 10X companies changed only 15 percent of their SMaC-recipe ingredients on average during our observation period.

CHANGE IN SMaC-RECIPE ELEMENTS
(10X COMPANIES)

Company	Number of Elements	Number of Elements Changed (%)*	Number of Years to Change Elements	Average Number of Years to Change Elements	Number of Years to Make First Change
Amgen	10	1 (10%)	10	10	10
Biomet	12	1 (10%)	8	8	8
Intel	11	2 (20%)	23, 30	26	23
Microsoft	13	2 (15%)	21, 24	22	21
Progressive	9	2 (20%)	35, 40	37	35
Southwest Airlines	10	2 (20%)	23, 26	24	23
Stryker	9	1 (10%)	19	19	19
Average	10	15%	24		20

* Percentages are rounded off because numbers of SMaC-recipe elements are approximate.

Finding 4. The comparison companies changed their SMaC-recipe elements more than the 10X companies (strong evidence). As shown in the following table, the comparison companies changed 60 percent of their SMaC-recipe elements on average—a far higher percentage than the 10X companies (15 percent).

CHANGE IN SMaC-RECIPE ELEMENTS
(COMPARISON COMPANIES)

Company	Number of Elements	Number of Elements Changed (%)*	Number of Years to Change Elements	Average Number of Years to Change Elements	Number of Years to Make First Change
Genentech	8	5 (60%)	14, 19,19,19,19	18	14
Kirschner	no SMaC recipe				
AMD	6	4 (65%)	15,15,15,29	18	15
Apple	8	5 (60%)	7,8,10,15,15	11	7

(continued on next page)

Company	Number of Elements	Number of Elements Changed (%)*	Number of Years to Change Elements	Average Number of Years to Change Elements	Number of Years to Make First Change
Safeco	7	5 (70%)	No info		
PSA	7	5 (70%)	16, 20, 26, 26 (1 no info)	22	16
USSC	7	4 (55%)	23, 29, 29, 31	28	23
Average	7	60%	19		15

* Percentages are rounded off because numbers of SMaC-recipe elements are approximate.

Finding 5. The 10X and comparison companies on average took a long time to change elements of their SMaC recipes (strong evidence). As the previous two tables illustrate, it took the 10X companies an average of 24 years to change an element (19 years for the comparisons). The 10X companies made the first change to their SMaC recipes after 20 years on average (the comparisons took 15 years).

RESEARCH FOUNDATIONS: LUCK ANALYSIS

The discussion in Chapter 7 is based on our analysis of 230 luck events. We analyzed the 10X and comparison companies' luck events (good and bad) to explore whether the companies experienced different magnitudes, types, and time distributions of luck events.[20]

OPERATIONAL DEFINITION OF LUCK. We defined a luck event as one where (1) some significant aspect of the event occurs largely or entirely independently of the actions of the key actors in the enterprise, (2) the event has a potentially significant consequence (good or bad) for the enterprise, and (3) the event has some element of unpredictability. There are two gradations of luck:

1. "Pure" luck, in which the occurrence of the event is completely independent of the actions of the key actors in the enterprise.
2. "Partial" luck, in which the occurrence of the event is largely but not completely independent of the actions of the key actors in the enterprise. To qualify as partial luck, some significant aspect of the event could not have been altered (prevented or caused) by the key actors, regardless of their skill.

In coding a luck event, it was important to pinpoint exactly what was the luck part of the event. To illustrate, consider Genentech in 1977.

That year the company was the first to accomplish gene splicing. That feat by itself was likely due to skill, not luck. But they were lucky that no one else had done this before them (an event outside their control, as they could not affect what others were doing). We coded the event "first to accomplish gene splicing" as "partial" luck (a combination of skill and luck).

In considering whether to classify an event as "good" luck or "bad" luck, the main consideration was how a reasonable person would have viewed the event *at the time when it happened*. We coded luck as good or bad based on this principle and not on the basis of later outcomes.

We systematically examined our company-documents and coded luck events by applying our definition and using the following categories:

- Pure luck, "Pure" (good or bad).
- Partial luck, "Partial" (good or bad).
- Medium importance, "Medium." The event had some impact on the success of the company (good or bad).
- High importance, "High." The event had a major impact on the success of the company (good or bad).

After each one of us (Jim and Morten) had *independently* completed a company pair, we compared notes and discussed discrepancies in our coding of luck events (they occurred for 5 percent of the events, indicating a high inter-rater reliability) and resolved them in follow-up meetings. This process yielded 230 luck events across all the companies in our data set (see examples from Amgen and Genentech in Chapter 7).

Finding 1. Both the 10X and comparison companies experienced good luck during our observation period (strong evidence). As shown in the following table, the 10X and comparison companies experienced on average seven and eight good-luck events, respectively.

GOOD-LUCK EVENTS

Matched Pairs	Number of Years Coded*		Number of Good-Luck Events		Number of Good-Luck Events per Decade†		10X Had More Good-Luck Events?
	10X	COMP‡	10X	COMP‡	10X	COMP‡	
Amgen and Genentech	23	27	10	18	4.3	6.7	Fewer
Biomet and Kirschner	26	9	4	4	1.5	4.4	Fewer
Intel and AMD	35	34	7	8	2.0	2.4	Similar
Microsoft and Apple	28	27	15	14	5.4	5.2	Similar
Progressive and Safeco	32	32	3	1	0.9	0.3	More
Southwest Airlines and PSA	36	43	8	6	2.2	1.4	More
Stryker and USSC	26	31	2	5	0.8	1.6	Fewer
Average	29.4	29.0	7.0	8.0	2.4	3.1	Similar/ Fewer
Total	206	203	49	56			

N=105 good-luck events

* From company founding to 2002. Progressive and Safeco coded from 1971, and Stryker from 1977 due to incomplete information.

† Controls for differences in number of years of observation within company pairs (e.g., for Amgen, 10 good-luck events divided by 2.3 decades).

‡ COMP = comparison companies.

Finding 2. The 10X companies did *not* experience substantially more *good-luck events* than the comparison companies (strong evidence). As summarized in the last column of the table above, there was no clear pattern. The 10X companies had more good-luck events than their comparisons in two pairs, fewer in three pairs, with two pairs being similar.

Finding 3. The 10X companies did *not* experience more high-importance and pure good-luck events than the comparison com-

panies (strong evidence). The table below reveals no substantial difference between the two groups for these important luck events; the 10X and comparison companies had a total of 36 and 40 such luck events, respectively.

GOOD-LUCK EVENTS BY TYPE

Type of Good-Luck Event	**10X**	**COMP***	**Ratio 10X/ COMP**	**10X Had More Good-Luck Events?**
Number of Good-Luck Events	49	56	0.9	
Number of High-Importance Good-Luck Events	22	28	0.8	Fewer
Number of Medium-Importance Good-Luck Events	27	28	1.0	
Number of Pure Good-Luck Events	14	12	1.2	Slightly more
Number of Partial Good-Luck Events	35	44	0.8	
SUM: Number of High-Importance *or* Number of Pure Good-Luck Events	36	40	0.9	Slightly fewer

N=105 good-luck events

* COMP = comparison companies.

Finding 4. The 10X companies did *not* experience substantially more *good-luck events* than the comparison companies *during their early years* (strong evidence). We performed this analysis to check whether either the 10X or comparison companies were luckier early on in their lives, but this was not the case (see the following table).

GOOD-LUCK EVENTS BY TYPE,
FOUNDING TO 5 AND 10 YEARS AFTERWARD

	10X	COMP*	10X Had More Good-Luck Events?
Average Number of Good-Luck Events from Founding to 5 Years Afterward	2.8	2.8	Same
Average Number of Good-Luck Events from Founding to 10 Years Afterward	5.0	4.5	Slightly more
Average Number of High-Importance Good-Luck Events from Founding to 5 Years Afterward	1.4	1.5	Slightly fewer
Average Number of High-Importance Good-Luck Events from Founding to 10 Years Afterward	2.8	2.3	Slightly more

Note: Two 10X companies (Stryker and Progressive) and one comparison company (Safeco) were excluded from this analysis because of lack of data for the early years after their founding.

* COMP = comparison companies.

Finding 5. The comparison companies did *not* experience substantially more *bad-luck events* than the 10X companies (strong evidence). It is possible that bad luck might explain why the comparison companies did not do as well. As the following table shows, however, the 10X and comparison companies experienced about the same number of bad-luck events (9.3 and 8.6 on average, respectively).

BAD-LUCK EVENTS

Matched Pairs	Number of Years Coded*		Number of Bad-Luck Events		Number of Bad-Luck Events per Decade†		Comparison Had More Bad-Luck Events?
	10X	COMP‡	10X	COMP	10X	COMP	
Amgen and Genentech	23	27	9	9	3.9	3.3	Fewer
Biomet and Kirschner	26	9	7	4	2.7	4.4	More
Intel and AMD	35	34	14	11	4.0	3.2	Fewer
Microsoft and Apple	28	27	9	7	3.2	2.6	Fewer
Progressive and Safeco	32	32	8	10	2.5	3.1	More
Southwest Airlines and PSA	36	43	13	13	3.6	3.0	Fewer
Stryker and USSC	26	31	5	6	1.9	1.9	Same
Average	29.4	29.0	9.3	8.6	3.2	3.1	Similar
Total	206	203	65	60			

N=125 bad-luck events

* From founding to 2002. Progressive and Safeco coded from 1971, and Stryker from 1977 due to incomplete information.

† Controls for differences in number of years of observation within company pairs.

‡ COMP = comparison companies.

Finding 6. The comparison companies did *not* experience substantially more *bad-luck events* than the 10X companies *during their early years* (strong evidence). It's possible that the comparison companies did less well because they experienced more bad luck initially, but this was not the case, as shown in the following table.

BAD-LUCK EVENTS BY TYPE, FOUNDING TO 5 OR 10 YEARS AFTERWARD

	10X	COMP*	Comparison Had More Bad-Luck Events?
Average Number of Bad-Luck Events from Founding to 5 Years Afterward	1.2	0.8	Slightly fewer
Average Number of Bad-Luck Events from Founding to 10 Years Afterward	3.0	1.7	Fewer
Average Number of High-Importance Bad-Luck Events from Founding to 5 Years Afterward	0.2	0	Similar
Average Number of High-Importance Bad-Luck Events from Founding to 10 Years Afterward	0.6	0.2	Slightly fewer

Note: Two 10X companies (Stryker and Progressive) and one comparison company (Safeco) were excluded from this analysis because of lack of data for the early years after their founding.

* COMP = comparison companies.

RESEARCH FOUNDATIONS: HOCKEY HALL OF FAME ANALYSIS

As discussed in Chapter 7, we compared the distribution of birth months in the Canadian general population with that of the truly great Canadian-born hockey players—those inducted into the Hockey Hall of Fame.

We performed the following analysis, with the assistance of research associate Lorilee Linfield. We first collected birth-month data for Canadian-born Hockey Hall of Fame inductees who were born between 1950 and 1966, and who had played at least one season in the National Hockey League (NHL).[21] We focused on players born 1950 or later to ensure data reliability and to perform the analysis in the most recent era. (In a follow-up analysis, we went back to 1873, used a larger sample, and came to the same conclusion.)[22]

We then collected birth-month data for the Canadian general population from 1951 to 1966 and tabulated these by months, quarters, and half-years.[23]

Finding 1. There is no disproportionate number of Hockey Hall of Fame inductees born in Canada between January and March (strong evidence). If anything, there might be a slightly disproportionate number born from October to December (1.9 percent more than in the general population), although the numbers are too small to draw any conclusion except that there is no meaningful difference across the birth-month groups.

BIRTH-MONTH DISTRIBUTIONS FOR THE CANADIAN-BORN HOCKEY HALL OF FAME INDUCTEES WHO PLAYED IN THE NHL AND FOR THE CANADIAN POPULATION

Birth-Month Range	Canadian-Born Hockey Hall of Fame Inductees (%)*	Canadian Population (%)	Hockey Hall of Fame percentage Minus Canadian-Population percentage (%)
January–March	22.9%	24.4%	–1.5%
April–June	25.7%	26.1%	–0.4%
July–September	25.7%	25.7%	0%
October–December	25.7%	23.8%	1.9%
January–June	48.6%	50.5%	–1.9%
July–December	51.4%	49.5%	1.9%

* N=35 (We also increased the sample size to 155 by going further back in time and came to the same conclusion.) Data is through 2009 Induction Year.

NOTES

CHAPTER 1: THRIVING IN UNCERTAINTY

1. Jason Zweig, "Risk-Management Pioneer and Best-Selling Author Never Stopped Insisting Future is Unknowable," *Wall Street Journal*, June 13, 2009, A14.
2. Source for all stock-return calculations in this work: ©200601 CRSP®, Center for Research in Security Prices. Booth School of Business, The University of Chicago. Used with permission. All rights reserved. http://www.crsp.chicagobooth.edu. Key definitions include:

    - *Monthly Total Return*: The total return to shareholders in a given month, including dividends reinvested, for an individual security.
    - *Cumulative Stock Return*: The compounded value of $Y invested in an individual security between times t1 and t2, using the formula $Y x (1 + Monthly Total Return @ m1) x (1 + Monthly Total Return @ m2) x . . . (1 + Monthly Total Return @ t2); where m1 = end of the first month following t1, m2 = end of the second month following t1, and so forth.
    - *General Stock Market* (also called *general market* or just the *market*): NYSE/AMEX/NASDAQ value-weighted return, which consists of the combined market value of all companies traded on these exchanges (including dividends reinvested), weighted by the capitalization of the company divided by the capitalization of the market.
    - *Cumulative Return Ratio to the Market*: At the end of any given time period, this ratio is calculated as the cumulative return of $Y invested in the company divided by the cumulative return of $Y invested in the general stock market, such that $Y is invested in both the company and the market on the same date.

3. "Southwest Airlines Co.: Presentation by Howard D. Putnam, President and Chief Executive Officer, Before the Dallas Association of In-

vestment Analysts," *Wall Street Transcript*, May 28, 1979; Jon Birger, "30-Year Super Stocks: Money Magazine Finds the Best Stocks of the Past 30 Years," *Money Magazine*, October 9, 2002; Southwest Airlines Co., *Fiscal 1976 Annual Report* (Dallas: Southwest Airlines Co., 1976). Source for all stock-return calculations in this work: ©200601 CRSP®, Center for Research in Security Prices. Booth School of Business, The University of Chicago. Used with permission. All rights reserved. http://www.crsp.chicagobooth.edu.

4. Invest in each company on December 31, 1972, and hold investment until December 31, 2002; if a company was not public on December 31, 1972, grow investment at general stock market rate of return until first month of CRSP data available for the company. Invest the same amount in the general market on December 31, 1972, and hold market investment until December 31, 2002. Divide cumulative value of company by cumulative value of market on December 31, 2002.
5. Regarding Industry Indices: We constructed an industry index for each company in the study, using Standard Industrial Classification (SIC) codes. The SIC code for a company may change over time; if there was only one SIC code since the company's IPO date, we used that code for the index; if there was more than one SIC code across the era, we used all relevant SIC codes to create one index. The data in the table show the value of making an investment in each of the 10X cases at the end of the month they each first appeared in the CRSP database held through December 31, 2002, divided by the value of an investment of the same amount in each company's corresponding industry index across the same time period. To provide a clean comparison, the 10X cases and the comparison cases are not included in the indices for the calculations in this table.
6. "John Wooden: A Coaching Legend October 14, 1910–June 4, 2010," *UCLA Official Athletic Site*, http://www.uclabruins.com/sports/m-baskbl/spec-rel/ucla-wooden-page.html.
7. The chart corresponding to this calculation was created using the following methodology: Invest $1 split evenly across all seven 10X companies (10X portfolio) on December 31, 1972; also invest $1 in the general market. Calculate the cumulative value of the 10X portfolio and of the general market from December 31, 1972, through December 31, 2002. Each month, calculate the ratio of the cumulative value-to-date of the 10X portfolio to the cumulative value-to-date of the general market. For months in which CRSP data is not available for a specific company (usually due to a company's not yet being publicly traded, merged, or acquired), use the general market returns in lieu of company returns. Repeat the same process for a portfolio of the seven comparison companies relative to the general market. Source for all stock-return calculations in this work: ©200601 CRSP®, Center for Research in Security Prices. Booth School of Business, The University of Chicago. Used

with permission. All rights reserved. http://www.crsp.chicagobooth.edu.

8. Jim Carlton, "Apple Computer is for Sale, But Buyers Prove Elusive," *Wall Street Journal*, January 19, 1996, B2. Source for all stock-return calculations in this work: ©200601 CRSP®, Center for Research in Security Prices. Booth School of Business, The University of Chicago. Used with permission. All rights reserved. www.crsp.chicagobooth.edu.
9. The companies in these studies included 3M, A&P, Abbott, Addressograph, AMD, American Express, Ames, Amgen, Apple, Bank of America, Best Buy, Bethlehem Steel, Biomet, Boeing, Bristol Myers/Squibb, Burroughs, Chase Manhattan, Chrysler, Circuit City, Citicorp, Colgate, Columbia Pictures, Eckerd, Fannie Mae, Ford, Genentech, General Electric, General Motors, Gillette, Great Western, Harris, Hasbro, Hewlett-Packard, Howard Johnson, IBM, Intel, Johnson & Johnson, Kenwood, Kimberly-Clark, Kirschner, Kroger, Marriott, McDonnell Douglas, Melville, Merck, Microsoft, Motorola, Nordstrom, Norton, Nucor, Pacific Southwest Airlines, Pfizer, Philip Morris, Pitney Bowes, Procter & Gamble, Progressive, R. J. Reynolds, Rubbermaid, Safeco, Scott Paper, Silo, Sony, Southwest Airlines, Stryker, Teledyne, Texas Instruments, Upjohn, USSC, Walgreens, Wal-Mart, Walt Disney, Warner Lambert, Wells Fargo, Westinghouse, and Zenith.
10. "Quotes from the Past," *Create the Future*, http://www.createthefuture.com/past_quotes.htm.

CHAPTER 2: 10XERS

1. Roald Amundsen, *The South Pole* (McLean, VA: IndyPublish.com, 2009), 192.
2. Roald Amundsen, *The South Pole* (McLean, VA: IndyPublish.com, 2009), "First Account," 204–5, 209; Roland Huntford, *The Last Place on Earth* (New York: Modern Library, 1999), 3, 11, 49, 109, 143, 167, 187, 204, 291, 371, 378, 400, 402–3, 433, 445, 468, 477, 490, 497, 506–7, 509, 516, 525–26.
3. Roald Amundsen, *The South Pole* (McLean, VA: IndyPublish.com, 2009), 31, 264–69; Roland Huntford, *The Last Place on Earth* (New York: Modern Library, 1999), 67, 91–94, 97–98, 100–101, 124, 250, 256, 332, 334, 337, 340–41, 351, 400, 407, 416, 422, 443–46, 468–76, 488, 497–99, 516, 523–25, 537–38.
4. Roland Huntford, *The Last Place on Earth* (New York: Modern Library, 1999), 444.
5. Robert McGough, "Executive Critical of 'Managed' Earnings Doesn't Mind if the Street Criticizes Him," *Wall Street Journal*, April 16, 1999, C1; Christopher Oster, "After Reg FD, Progressive Sets Bold Move," *Wall Street Journal*, May 11, 2001, C1; *Yahoo! Finance*, http://www.finance.yahoo.com.

6. Robert McGough, "Executive Critical of 'Managed' Earnings Doesn't Mind if the Street Criticizes Him," *Wall Street Journal*, April 16, 1999, C1.
7. "Progressive Debuts Monthly Financial Information," A. M. *Best Newswire*, May 18, 2001; Amy Hutton and James Weber, "Progressive Insurance: Disclosure Strategy," *Harvard Business School*, case study #9–102–012 (Boston: Harvard Business School Publishing, 2001), 7, 10.
8. Katrina Brooker, Herb Kelleher, and Reporter Associate Alynda Wheat, "The Chairman of the Board Looks Back," *Fortune*, May 21, 2008, 4; Charles O'Reilly and Jeffrey Pfeffer, "Southwest Airlines: Using Human Resources for Competitive Advantage (A)," *Graduate School of Business, Stanford University* case study #HR–1A (Palo Alto, CA: Graduate School of Business, Stanford University, 2003), 6.
9. "Herb and his Airline," *60 Minutes*, CBS, May 27, 1990, Television; Jane Gibson, "Work Hard, Play Hard," *Smart Business*, November 2005.
10. Jan Jarboe Russell, "A Boy and His Airline," *Texas Monthly*, April 1989.
11. Tonda Montague (Ed.), Employee Communications, *Southwest Airlines: 30 Years. One Mission. Low Fares.* Supplement to Southwest Airlines Co., *Fiscal 2001 Annual Report* (Dallas: Southwest Airlines Co., 2001), 35.
12. "Southwest Airlines Co. (LUV)," *Wall Street Transcript*, June 8, 1987.
13. John Kirkpatrick, "Clownish in Public, Southwest Airlines Executive can be Ruthless to Rivals," *Knight Ridder/Tribune Business News*, March 20, 2001.
14. Andy Grove with Bethany McLean, "Taking on Prostate Cancer," *Fortune*, May 13, 1996.
15. Andy Grove with Bethany McLean, "Taking on Prostate Cancer," *Fortune*, May 13, 1996.
16. Robert B. Cialdini and Noah J. Goldstein, "Social Influence: Compliance and Conformity," *Annual Review of Psychology*, February 2004, 591–621.
17. Roald Amundsen, *The South Pole* (McLean, VA: IndyPublish.com, 2009), 29; Roland Huntford, *The Last Place on Earth* (New York: Modern Library, 1999), 241–43.
18. Bro Uttal, "Inside the Deal that Made Bill Gates $350,000,000," *Fortune*, July 21, 1986, 27.
19. Walter Isaacson, "In Search of the Real Bill Gates," *Time*, January 13, 1997. (This article also appeared in the October 20, 2005, edition of *Time*.)
20. "For Bill Gates, Micros are Personal," *Information Week*, August 14, 1989; "The Bill Gates Interview," *Playboy*, 1994; Brent Schlender, "What Bill Gates Really Wants," *Fortune*, January 16, 1995, 34.
21. James Wallace and Jim Erickson, *Hard Drive* (New York: HarperBusiness, 1992), 402–3; Lee Gomes, "Microsoft's Gates Eyes Challenges," *San Jose Mercury News*, June 18, 1991, 1C; Lee Gomes, "Candid Memo

Costs Microsoft's Gates a Fortune," *San Jose Mercury News*, June 20, 1991, 1F; Rich Karlgaard, "ASAP Interview: Bill Gates," *Forbes*, December 7, 1992; *Yahoo! Finance*, http://www.finance.yahoo.com.

22. Kathy Rebello and John Hillkirk, "Sculley to Take a Break; Sabbaticals at Core of Apple Perks," *USA Today*, June 10, 1988, 01B.
23. Kathy Rebello and John Hillkirk, "Sculley to Take a Break; Sabbaticals at Core of Apple Perks," *USA Today*, June 10, 1988, 01B.
24. John Markoff, "Visionary Apple Chairman Moves On," *New York Times*, October 16, 1993; Chris Higson and Tom Albrighton, "Apple Computer's Financial Performance," *London Business School*, case study #CS 08–012 (London: London Business School Publishing, 2008), 8, 11; Apple Inc., *Fiscal 1994 and 1996 10-Ks* (Cupertino, CA: Apple Inc., 1994 and 1996).
25. Johanna M. Hurstak and David B. Yoffie, "Reshaping Apple Computer's Destiny 1992," *Harvard Business School*, case study #9–393–011 (Boston: Harvard Business School Publishing, 1992), 9.
26. Mead Jennings, "Staying the Course," *Airline Business*, February 1992, 52; Elizabeth Corcoran, "Intel's Blunt Edge," *Washington Post*, September 8, 1996, H01; Arlene Weintraub and Amy Barrett, "Amgen: Up from Biotech," *Business Week*, March 18, 2002, 70; Lee Gomes, "Microsoft's Gates Eyes Challenges," *San Jose Mercury News*, June 18, 1991, 1C.
27. Geoffrey Smith and James Ellis, "Pay That was Justified—And Pay that Just Mystified," *Business Week*, May 6, 1991; James Ellis, "You Don't Necessarily Get What You Pay For," *Business Week*, May 4, 1992, 144; "CEO/Company Interview: Dr. Dane A. Miller, Biomet, Inc.," *Wall Street Transcript*, December 2000; Steve Kaelble, "Money's Worth: Which CEOs Deliver the Best Return?" *Indiana Business Magazine*, July 1, 1999, 15; Matthew Herper, "Dane Miller: CEO Value to the Bone," *Forbes*, May 8, 2001; Tom Schuman, "Biomet and CEO Dane Miller," *CEO Magazine*, November/December 2002, 44.
28. James Ellis, "You Don't Necessarily Get What You Pay For," *Business Week*, May 4, 1992; "CEO/Company Interview: Dr. Dane A. Miller, Biomet, Inc.," *Wall Street Transcript*, December 2000.
29. Stephen Phillips, "Driven to Succeed: Peter Lewis, Progressive's Artful Chief Exec, Aims to Overtake Auto Insurance Industry's Leaders," *Plain Dealer*, September 1, 1996, 1.I; Gregory David, "Chastened?" *Financial World*, January 4, 1994, 39; Carol J. Loomis, "Sex. Reefer? And Auto Insurance!" *Fortune*, August 7, 1995, 76; personal conversation with author.
30. Mike Casey, "Insurer Favors Low-Risk Route; Progressive Corp.'s Personnel Help Write a Policy for Success," *Crain's Cleveland Business*, February 23, 1987, 2; Mike Casey, "Insurer Favors Low-Risk Route; Progressive Corp.'s Personnel Help Write a Policy for Success," *Crain's Cleveland Business*, February 23, 1987, 2; The Progressive Corporation, *Fiscal 1991 Annual Report* (Mayfield Heights, OH: The Progres-

sive Corporation, 1991), 14; The Progressive Corporation, *Fiscal 1996 Annual Report* (Mayfield Village, OH: The Progressive Corporation, 1996), 30.

31. Carol J. Loomis, "Sex. Reefer? And Auto Insurance!" *Fortune*, August 7, 1995, 76.
32. Andrew Bary, "No. 4 Progressive Closes In On Auto Insurance Leaders," *Wall Street Journal*, November 8, 2009.
33. "Intel Executive Biography: Gordon Moore," *Intel Corporation*, http://www.intel.com/; Gene Bylinsky, "How Intel Won Its Bet on Memory Chips," *Fortune*, November 1973; Leslie Berlin, *The Man Behind the Microchip* (New York: Oxford University Press, 2005), 244.
34. Bro Uttal, "Inside The Deal That Made Bill Gates $350,000,000," *Fortune*, July 21, 1986.
35. Bill Gates, "Microsoft's Bill Gates: Harvard Commencement Speech Transcript," *Network World*, June 8, 2007; Ruthie Ackerman, "Gates Fights To Eradicate Malaria," *Forbes*, October 19, 2007; "The Meaning of Bill Gates," *Economist*, June 26, 2008; "Microsoft's Tradition of Innovation," *Microsoft Corporation*, October 25, 2002, http://www.microsoft.com/about/companyinformation/ourbusinesses/profile.mspx.
36. Michael A. Verespej, "Recession? What Recession? Southern Gentleman John Brown Achieves 20 Percent Earnings Growth Annually—No Matter What," *Chief Executive*, June 2002, 45; "John W. Brown," *Michigan Economic Development Corporation*, http://www.themedc.org/Executive-Committee/John-Brown/.
37. Michael Hiestand, "Flying the Wacky Skies with Southwest's CEO," *Adweek's Marketing Week*, July 10, 1989, 31; Charles O'Reilly and Jeffrey Pfeffer, "Southwest Airlines: Using Human Resources for Competitive Advantage (A)," *Graduate School of Business, Stanford University*, case study #HR–1A (Palo Alto, CA: Graduate School of Business, Stanford University, 2003), 6; "Southwest Airlines Company (LUV), *Wall Street Transcript*, June 5, 1989; "Officer Biographies: Herbert D. Kelleher," *Southwest Airlines*, http://www.southwest.com/swamedia/bios/herb_kelleher.html.
38. Steven Litt, "This Lone Ranger has Nothing to Hide," *Plain Dealer*, September 29, 2002, A1; April Dougal Gasbarre (updated by David Bianco), "The Progressive Corporation," *International Directory of Company Histories* (New York: St. James Press, 1999), 396.
39. Geoffrey Smith, "The Guts to Say 'I Was Wrong,' " *Forbes*, May 28, 1979.
40. "Jerry Sanders's Act is Cleaning Up," *Fortune*, October 15, 1984; Jeffrey L. Rodengen, *The Spirit of AMD: Advanced Micro Devices* (Fort Lauderdale, FL: Write Stuff Enterprises Inc., 1998), 22–24.
41. Tonda Montague (Ed.), Employee Communications, *Southwest Airlines: 30 Years. One Mission. Low Fares.* Supplement to Southwest Airlines Co., *Fiscal 2001 Annual Report* (Dallas: Southwest Airlines Co., 2001),

35; Seanna Browder, "How George Rathmann Mastered the Science of the Deal," *Business Week*, November 30, 1998; Arthur Kornberg, *The Golden Helix* (Sausalito, CA: University Science Books, 1995), 205.

CHAPTER 3: 20 MILE MARCH

1. From the poem "A Blessing" by Mekeel McBride, reproduced in Ted Kooser, *The Poetry Home Repair Manual* (Lincoln, NE: University of Nebraska Press, 2005), 141.
2. Stryker Corporation, *Fiscal 1980, 1982, 1984, 1986, 1988, 1990, 1992, 1994, 1996, and 1997 Annual Reports* (Kalamazoo, MI: Stryker Corporation, 1980, 1982, 1984, 1986, 1988, 1990, 1992, 1994, 1996, and 1997); United States Surgical Corporation, *Fiscal 1979–2002 Annual Reports* (Norwalk, CT: United States Surgical Corporation, 1979–2002). When calculating annual growth in cases in which earnings were negative, we used the formula (Year 2–Year 1)/Absolute Value (Year 1). Stryker made note of a 1990 extraordinary gain, which it did not count in its year-to-year comparisons of net income; if this extraordinary gain is removed from the calculations, the standard deviation drops to 7 percentage points. By "never posted a net loss," we are referring to the fact that Stryker never had net income below zero during this time.
3. Source for all stock-return calculations in this work: ©200601 CRSP®, Center for Research in Security Prices. Booth School of Business, The University of Chicago. Used with permission. All rights reserved. http://www.crsp.chicagobooth.edu; Laura M. Holson, "Tyco to Pay $3.3 Billion in Stock for U.S. Surgical," *New York Times*, May 26, 1998.
4. Stryker Corporation, *Fiscal 1979 Annual Report* (Kalamazoo, MI: Stryker Corporation, 1979); Zina Sawaya, "Focus through Decentralization," *Forbes*, November 11, 1991, 242; Michael A. Verespej, "Recession? What Recession? Southern Gentleman John Brown Achieves 20 Percent Earnings Growth Annually—No Matter What," *Chief Executive*, June 2002, 45; "John W. Brown Profile," *Forbes*, http://people.forbes.com/profile/john-w-brown/35968; Eric Whisenhunt, "Stryker Force: Divide, Conquer, and Be First with the New," *Michigan Business Magazine*, November 1985, 36.
5. Geoffrey Brewer, "20 Percent—Or Else!" *Sales and Marketing Management*, November 1994, 66; Matt Walsh, "Avoiding the Snorkel Award," *Forbes*, January 2, 1995, 180.
6. Geoffrey Brewer, "20 Percent—Or Else!" *Sales and Marketing Management*, November 1994, 66.
7. Michael A. Verespej, "Recession? What Recession? Southern Gentleman John Brown Achieves 20 Percent Earnings Growth Annually—No Matter What," *Chief Executive*, June 2002, 45.
8. Geoffrey Brewer, "20 Percent—Or Else!" *Sales and Marketing Management*, November 1994, 66; Steve Watkins, "Stryker Corp./Kalamazoo,

Michigan: Failure Not An Option For This Manufacturer," *Investor's Business Daily*, September 25, 2001, A10.

9. Stryker Corporation, *Fiscal 1979–2002 Annual Reports* (Kalamazoo, MI: Stryker Corporation, 1979–2002) (The *Fiscal 1997 Annual Report* [page 2] states, "Stryker Corporation achieved its net earnings goal in 1997, delivering our 21st consecutive year of 20% growth or better." Stryker made note of a 1990 extraordinary gain, which it did not count in its year-to-year comparisons of net income. For calculating Stryker's frequency of hitting the 20 Mile March, we used the net earnings chart on page 2 of the *Fiscal 2002 Annual Report*.); "Corporate Critic's Confidential," *Wall Street Transcript*, February 27, 1989.
10. Edward A. Wyatt, "Just What the Doctor Ordered," *Barron's*, June 4, 1990; Ron Winslow, "Heard on the Street: U.S. Surgical Shares Plunge: Is Fall Over?" *Wall Street Journal*, April 9, 1993, C1; Christopher Tucher, "Now, Lee Hirsch Wants to Sew Up Sutures," *Business Week*, August 7, 1989, 74–75; "FDA Will Let Stand Its Decision to Speed Approval of Sutures," *Wall Street Journal*, August 25, 1989, 1; United States Surgical Corporation, *Fiscal 1988 Annual Report* (Norwalk, CT: United States Surgical Corporation, 1988).
11. Ron Winslow, "Heard on the Street: U.S. Surgical Shares Plunge: Is Fall Over?" *Wall Street Journal*, April 9, 1993, C1; Felicia Paik, "Unhealthy Sales Afflict Many Suppliers of Medical Goods as Hospitals Cut Costs," *Wall Street Journal*, January 14, 1994, B4B; Ron Winslow, "As Marketplace Shifts, U.S. Surgical Needs Patching Up," *Wall Street Journal*, February 18, 1994, B4; "Recent Suture Prices Sliding Downward, as Hospital Buyers Cast Votes for Ethicon," *Hospital Materials Management*, August 1996, 1; Howard Rudnitsky, "On the Mend," *Forbes*, December 2, 1996, 58; "Shareholders Approve Tyco's Acquisition of U.S. Surgical Corporation," *PR Newswire*, October 1, 1998; Laura M. Holson, "Tyco to Pay $3.3 Billion in Stock for U.S. Surgical," *New York Times*, May 26, 1998; United States Surgical Corporation, *Fiscal 1989, 1991, 1992, 1995, and 1997 Annual Reports* (Norwalk, CT: United States Surgical Corporation, 1989, 1991, 1992, 1995, and 1997).
12. Richard M. McCabe, "Airline Industry Key Success Factors," *Graziadio Business Report* (Malibu, CA: Pepperdine University, 2006); Howard D. Putnam with Gene Busnar, *The Winds of Turbulence* (Reno, NV: Howard D. Putnam Enterprises Inc., 1991), 83; Southwest Airlines Co., *Fiscal 2001 Proxy Statement* (Dallas: Southwest Airlines Co., 2001); Southwest Airlines Co., *Fiscal 2002 Annual Report* (Dallas: Southwest Airlines Co., 2002).
13. "Southwest Airlines Co.," *Wall Street Transcript*, May 28, 1979; "Texas Gets Bigger," *Forbes*, November 12, 1979, 88–89; Charles O'Reilly and Jeffrey Pfeffer, "Southwest Airlines: Using Human Resources for Competitive Advantage (A)," *Graduate School of Business, Stanford University*, case study #HR–1A (Palo Alto, CA: Graduate School of Busi-

ness, Stanford University, 1995), 8; Tonda Montague (Ed.), Employee Communications, *Southwest Airlines: 30 Years. Mission. Low Fares,* (Supplement to Southwest Airlines Co. *Fiscal 2001 Annual Report*), 2001; Southwest Airlines Co., *Fiscal 1996 and 2001 Annual Reports* (Dallas: Southwest Airlines Co., 1996 and 2001).

14. Southwest Airlines Co., *Fiscal 2001 Proxy Statement* (Dallas: Southwest Airlines Co., 2001); Southwest Airlines Co., *Fiscal 2002 Annual Report* (Dallas: Southwest Airlines Co., 2002); data for chart comes from sources in this and the prior two end notes.

15. "The Progressive Corporation," *Wall Street Transcript,* February 28, 1972; Thomas A. King, "The Progressive Corporation (PGR)," *Wall Street Transcript,* January 14, 2002; The Progressive Corporation, *Fiscal 1971 Annual Report* (Cleveland, OH: The Progressive Corporation, 1971); The Progressive Corporation, *Fiscal 1976 Annual Report* (Mayfield Village, OH: The Progressive Corporation, 1976).

16. Peter B. Lewis, "The Progressive Corporation: Address to the New York Society of Security Analysts," *Wall Street Transcript,* February 28, 1972; Elisabeth Boone, "Recipe for Success," *Rough Notes,* April 2002, 42; The Progressive Corporation, *Fiscal 1971 Annual Report* (Cleveland, OH: The Progressive Corporation, 1971); The Progressive Corporation, *Fiscal 1976, 1986, 1996, 2001, and 2003 Annual Reports* (Mayfield Village, OH: The Progressive Corporation, 1976, 1986, 1996, 2001, and 2003).

17. Eric Whisenhunt, "Stryker Force: Divide, Conquer, and Be First with the New," *Michigan Business Magazine,* November 1985, 36; Mike Casey, "Insurer Favors Low-Risk Route; Progressive Corp.'s Personnel Help Write a Policy for Success," *Crain's Cleveland Business,* February 23, 1987, 2; Noreen Seebacher, "Stryker Products: Just What the Doctor Ordered," *Detroit News,* May 6, 1991, 3F; Nicolaj Siggelkow and Michael E. Porter, "Progressive Corporation," *Harvard Business School,* case study #9–797–109 (Boston: Harvard Business School Publishing, 1998); Elisabeth Boone, "Recipe for Success," *Rough Notes,* April 2002, 42; The Progressive Corporation, *Fiscal 1981 Annual Report* (Mayfield Village, OH: The Progressive Corporation, 1981), 11.

18. W. L. Campbell, "General of America Earnings Increase to New High Level," *National Underwriter,* February 7, 1964, 1; "Safeco Corporation," *Commercial and Financial Chronicle,* October 3, 1968; "Safeco Corporation," *Wall Street Transcript,* June 9, 1969; "Safeco Corporation," *Wall Street Transcript,* July 12, 1976; Art Garcia, "Spotlight on Safeco Corp.," *Journal of Commerce,* January 24, 1977, 2; "Safeco: 'Redlining' Two States to Bolster Insurance Profits," *Business Week,* July 17, 1979, 88; William Mehlman, "Safeco Continues to Stand Out in Depressed Casualty Group," *Insiders' Chronicle,* October 26, 1979, 7; "Safeco Corporation," *Wall Street Transcript,* August 8, 1983; "Safeco Reports Loss of $41m During the First Quarter of '85," *National Un-*

derwriter, May 3, 1985, 6; John Davies, "Safeco Profit Weakness Blamed on Junk Bonds," *Journal of Commerce*, April 30, 1990; Greg Heberlein, "Safeco Plea Seeks a Change of Shirt," *Seattle Times*, May 3, 1990, E2; "Safeco Corporation," *Wall Street Transcript*, July 12, 1976; Peter Neurath, "Safeco Loses Millions on Commercial, Auto Lines," *Puget Sound Business Journal*, March 19, 1990, 3.

19. Leslie Scism, "Safeco Plans $2.82 Billion Acquisition," *Wall Street Journal*, June 9, 1997, A3; Judy Greenwald, "SAFECO Bids $2.8 Billion for American States," *Business Insurance*, June 16, 1997, 1; Thomas A. McCoy, "Safeco's Huge Bet on the Independent Agency System," *Rough Notes*, December 1997, 34; Peter Neurath, "Safeco's Stodgy Image Changes with Latest Move," *Business Journal—Portland*, October 10, 1997, 29; Beth Neurath, "Fun is the Best Policy," *Puget Sound Business Journal*, December 25, 1998, 6; Boh A. Dickey, "CEO Interview with Boh A. Dickey—Safeco Corporation (SAFC)," *Wall Street Transcript*, April 27, 1999; Paula L. Stepankowsky, "After Revamp, Safeco's CEO is Focusing Energies on Most Profitable Operations," *Wall Street Journal*, March 27, 2002, B5C; Safeco Insurance Company of America, *Fiscal 1996 and 1997 Annual Reports* (Seattle, WA: Safeco Insurance Company of America, 1996 and 1997). Note: The "falling more than 60 percent behind the general stock market" is based on the $2.8 billion acquisition price divided by the shareholders' equity on the company's 1996 balance sheet.

20. Carol Tice, "Acquisition Put Safeco in a Long Slump," *Puget Sound Business Journal*, October 8, 1999; Khanh T. L. Tran, "Eigsti and Stoddard Are Leaving Safeco as Insurer Struggles to Regain Footing," *Wall Street Journal*, August 4, 2000, B5; Ruth Levine, "Safeco Rewrites Growth Policy," *Puget Sound Business Journal*, August 11, 2000, 1; "UPDATE: Safeco's Newly Named CEO Believes He's Prepared for Job," *A. M. Best Newswire*, February 2, 2001; Susanne Sclafane, "SAFECO Chooses Former CNA Exec for Chief Executive Spot," *National Underwriter*, February 5, 2001, 2; Safeco Insurance Company of America, *Fiscal 1986, 1991, and 1996–2003 Annual Reports* (Seattle, WA: Safeco Insurance Company of America, 1986, 1991, and 1996–2003). Source for all stock-return calculations in this work: ©200601 CRSP®, Center for Research in Security Prices. Booth School of Business, The University of Chicago. Used with permission. All rights reserved. http://www.crsp.chicagobooth.edu. Progressive's combined ratios cover only vehicle insurance, while Safeco's combined ratios cover vehicle insurance plus other insurance businesses under the property and casualty umbrella. Despite this difference, our point is valid because our analysis looked at whether each company met its *own* standard for an underwriting profit as it had defined it.

21. "Innovative Intel," *Economist*, June 16, 1979, 94; Michael Annibale, "Intel: The Microprocessor Champ Gambles on Another Leap Forward,"

Business Week, April 14, 1980, 98; Mimi Real and Robert Warren, *A Revolution in Progress . . . A History of Intel to Date* (Santa Clara, CA: Intel Corporation, 1984), 4; Gordon E. Moore, "Cramming More Components onto Integrated Circuits," *Proceedings of the IEEE*, January 1998, 82–83; Leslie Berlin, *The Man Behind the Microchip* (New York: Oxford University Press, 2005), 160; "Moore's Law," *Intel Corporation*, http://www.intel.com/technology/mooreslaw/.

22. See relevant notes from Stryker discussion earlier in chapter.
23. See relevant notes from USSC discussion earlier in chapter.
24. See relevant notes from Southwest discussion earlier in chapter.
25. "USAir Completes Takeover of Pacific Southwest," *Washington Post*, May 30, 1987.
26. See relevant notes from Progressive discussion earlier in chapter.
27. See relevant notes from Safeco discussion earlier in chapter.
28. "Moore's Law," *Intel Corporation*, http://www.intel.com/technology/mooreslaw/.
29. See relevant notes from AMD discussion later in chapter.
30. Stratford P. Sherman, "Microsoft's Drive to Dominate Software," *Fortune*, January 23, 1984, 82; Greg Heberlein, "Microsoft Stock Filing Unveils Secrets," *Seattle Times*, February 4, 1986, B1; James Wallace and Jim Erickson, *Hard Drive* (New York: HarperBusiness, 1992), 314; Brent Schlender, "What Bill Gates Really Wants," *Fortune*, January 16, 1995; Jim Carlton, *Apple* (New York: Random House, 1997), 132; Steve Hamm, "Gates on Bullies, Browsers—and the Future," *Business Week*, January 19, 1998; David Bank, "Paneful Struggle: How Microsoft's Ranks Wound up in Civil War over Windows' Future," *Wall Street Journal*, February 1, 1999, A1; Carl Johnston, Michael Rukstad, and David Yoffie, "Microsoft, 2000," *Harvard Business School*, case study #9–700–071 (Boston: Harvard Business School Publishing, 2000), 20–21.
31. Morgan Stanley & Co. and Hambrecht & Quist, "Prospectus: Apple Computer, Inc., Common Stock," *Apple Inc.*, December 12, 1980; John Eckhouse, "It's Final—Apple Chairman Resigns," *San Francisco Chronicle*, September 20, 1985; G. Pascal Zachary, "Apple Plans Cutbacks as its Profits Sour," *Wall Street Journal*, January 19, 1990, B1; Don Clark, "Apple's Gassee Confirms Resignation," *San Francisco Chronicle*, March 3, 1990, B2; Richard Brandt, "Information Processing: The Toughest Job in the Computer Business—Michael Spindler Tripled Apple Sales in Europe. Will his Magic Work in America?" *Business Week*, March 19, 1990, 118; Barbara Buell, "Apple: New Team, New Strategy," *Business Week*, October 15, 1990, 86; Bill Richards, Michael Gibbs, and Michael Beer, "Apple Computer (D): Epilogue," *Harvard Business School*, case study #9–492–013 (Boston: Harvard Business School Publishing, 1991), 3; "Apple Finance Chief Quits After Pushing for a Merger," *Wall Street Journal*, October 5, 1995, B1; Peter Burrows, "Almost Down to the Core? Apple is Facing a Disturbing Management Exodus," *Business Week*,

November 20, 1995; Kathy Rebello, "The Fall of an American Icon," *Business Week*, February 5, 1996; Jim Carlton and Lee Gomes, "Apple Computer Chief Amelio is Ousted," *Wall Street Journal*, July 10, 1997, A3; Jim Carlton, "Apple Names Steve Jobs Interim CEO," *Wall Street Journal*, September 17, 1997, A3; Apple Inc., *Fiscal 2002 10-K* (Cupertino, CA: Apple Inc., 2002).

32. James D. Berkley and Nitin Nohria, "Amgen Inc.: Planning the Unplannable," *Harvard Business School*, case study #9–492–052 (Boston: Harvard Business School Publishing, 1992), 11; Amy Tsao, "Amgen: Will Bigger be Better?" *Business Week*, January 2, 2002; David Stipp, "Biotech's New Colossus: Move Over, Big Pharma. Amgen Boasts Better Growth," *Fortune*, April 15, 2002; Frank DiLorenzo, "For Amgen, a Very Healthy Prognosis," *Business Week*, August 10, 2004; Amgen Inc., *Fiscal 1985, 1990, 1995 and 2000–2002 Annual Reports* (Thousand Oaks, CA: Amgen Inc, 1985, 1990, 1995, and 2000–2002).
33. See relevant notes from Genentech discussion later in chapter.
34. Geoffrey Smith, "Pay that was Justified—And Pay that Just Mystified," *Business Week*, May 6, 1991, 92; Michael Brush, "Millions in the Bank, if They Don't Stumble," *New York Times*, March 3, 1997, 3.6; "Dane Miller—Biomet Inc (BMET): CEO Interview," *Wall Street Transcript*, July 15, 2002; Biomet Inc., *Fiscal 1982, 1987, 1992, 1997, and 2002 Annual Reports* (Warsaw, IN: Biomet Inc., 1982, 1987, 1992, 1997, and 2002).
35. M. L. Mead: Scott & Stringfellow Inc., "Kirschner Medical Corporation—Company Report," *The Investext Group*, February 17, 1989; J. H. Berg: J. C. Bradford & Co., "Kirschner Medical Corporation—Company Report," *The Investext Group*, May 19, 1989, 1; Stan Hinden, "Kirschner Medical to Sell Surgical Lighting Division," *Washington Post*, April 3, 1990, D01; Jason Zweig, "The Bone Doctor's Plan," *Forbes*, January 20, 1992, 92; Jessica Hall, "Torn between Two Bidders: Kirschner's Enviable Dilemma," *Warfield's Business Record*, July 1, 1994, 3; Jessica Hall, "Kirschner Accepts Deal with Orthopedics Giant to End Seven-Week Bidding War," *Warfield's Business Record*, July 22, 1994, 11.
36. Mary Jo Waits, et al., *Beat the Odds* (Tempe and Phoenix, AZ: Morrison Institute for Public Policy and Center for the Future of Arizona, 2006); *Beat the Odds Institute*, http://www.beattheoddsinstitute.org/overview/index.php.
37. Mary Jo Waits, et al., *Beat the Odds* (Tempe and Phoenix, AZ: Morrison Institute for Public Policy and Center for the Future of Arizona, 2006), 16, 25, 29, 36, 43; *Arizona Indicators: A Program Managed by Morrison Institute for Public Policy*, http://arizonaindicators.org/education.
38. Louise Kehoe, "How Immodesty Becomes a Silicon Valley Resident," *Financial Times*, January 30, 1984, 10; "Advanced-Micro: Goal is to be No. 1 U.S. Integrated Circuit Producer by 1990," *Business Wire*, Septem-

ber 11, 1984; Peter Dworkin, "Silicon Valley's Vale of Tears," *U.S. News & World Report*, March 2, 1987, 47; Jeffrey L. Rodengen, *The Spirit of AMD: Advanced Micro Devices* (Ft. Lauderdale, FL: Write Stuff Enterprises Inc., 1998), 82–83; Moody's Investors Service and Mergent FIS Inc., *1973–1986 Moody's OTC Industrial Manual* (New York: Moody's Investors Service, 1973–86).

39. Advanced Micro Devices (AMD), *Fiscal 1987 Annual Report* (Sunnyvale, CA: Advanced Micro Devices, 1987).

40. Peter Dworkin, "Silicon Valley's Vale of Tears," *U.S. News & World Report*, March 2, 1987, 47; Dale Wettlaufer, "Interview with Vladi Catto," *Motley Fool*, June 21, 1996; Douglas A. Irwin, "Trade Policies and the Semiconductor Industry," *National Bureau of Economic Research*, January 1996, 27; Moody's Investors Service and Mergent FIS Inc., *1973–1986 Moody's OTC Industrial Manual* (New York: Moody's Investors Service, 1973–86). Source for all stock-return calculations in this work: ©200601 CRSP®, Center for Research in Security Prices. Booth School of Business, The University of Chicago. Used with permission. All rights reserved. http://www.crsp.chicagobooth.edu. Advanced Micro Devices, *Fiscal 1987 and 1997–2002 Annual Reports* (Sunnyvale, CA: Advanced Micro Devices, 1987 and 1997–2002).

41. Roald Amundsen, *The South Pole* (McLean, VA: IndyPublish.com, 2009), "The First Account," 213, 263; Roland Huntford, *The Last Place on Earth* (New York: Random House, 1999), 412–13, 419, 441–43, 466–67, 483–84.

42. "Hot Reception Seen Today for Genentech As First Gene-Splicing Firm to Go Public," *Wall Street Journal*, October 14, 1980, 6; Nell Henderson, "Biotech Breakthrough Focuses on Heart Attacks," *Washington Post*, October 12, 1986, H1; Charles McCoy, "Genentech's New CEO Seeks Clean Slate—Levinson Takes Charge At Biotech Firm After Raab's Ouster," *Wall Street Journal*, July 12, 1995, B6; Bernadette Tansey, "Genentech Proves the Skeptics Wrong," *San Francisco Chronicle*, December 21, 2003; Genentech Inc., *Fiscal 1985 and 1991 Annual Reports* (San Francisco: Genentech Inc., 1985 and 1991); *Genentech Inc.*, http://www.gene.com.

43. David R. Olmos, "Genentech Ousts CEO over Conflict Question," *Los Angeles Times*, July 11, 1985, D1; Charles McCoy, "Genentech's New CEO Seeks Clean Slate—Levinson Takes Charge At Biotech Firm After Raab's Ouster," *Wall Street Journal*, July 12, 1995, B6; Wayne Koberstein, "Youthful Maturity," *Pharmaceutical Executive*, March 1999, 47; "Arthur D. Levenson—Genentech Inc. (GNE) CEO Interview," *Wall Street Transcript*, January 26, 1998 (Note: Article title appears as original, with misspelling of Levinson's name.); net income data for chart comes from Genentech Inc., *Fiscal 1980–2008 Annual Reports* (San Francisco: Genentech Inc., 1980–2008). Source for all stock-return calculations in this work: ©200601 CRSP®, Center for Research in Secu-

rity Prices. Booth School of Business, The University of Chicago. Used with permission. All rights reserved. http://www.crsp.chicagobooth.edu; *Business Week Online,* http://investing.businessweek.com/businessweek/research/stocks/people/person.asp?personId=234085&ticker=DNA:CN.

44. "Arthur D. Levenson—Genentech Inc. (GNE) CEO Interview," *Wall Street Transcript*, January 26, 1998. (Note: Article title appears as original, with misspelling of Levinson's name.)

CHAPTER 4: FIRE BULLETS, THEN CANNONBALLS

1. Mimi Real and Robert Warren, *A Revolution in Progress . . . A History of Intel to Date* (Santa Clara, CA: Intel Corporate Communications Department, 1984), 17.
2. Gary Kissel, *Poor Sailors' Airline* (McLean, VA: Paladwr Press, 2002), viii, 21, 23, 69, 80, 116–17, 145, 171–72.
3. Gary Kissel, *Poor Sailors' Airline* (McLean, VA: Paladwr Press, 2002), 118–19; Richard Curry, "The Skies of Texas," *New York Times*, July 18, 1971; PSA Inc., *Fiscal 1967 Annual Report* (San Diego: PSA Inc., 1967).
4. "Love is Ammunition for a Texas Airline," *Business Week*, June 26, 1971; Gary Kissel, *Poor Sailors' Airline* (McLean, VA: Paladwr Press, 2002), 171.
5. Gary Kissel, *Poor Sailors' Airline* (McLean, VA: Paladwr Press, 2002), 171–72; Lamar Muse, *Southwest Passage: The Inside Story of Southwest Airlines' Formative Years* (Austin, TX: Eakin Press, 2002), 84; Christopher H. Lovelock, "Southwest Airlines (A)," *Harvard Business School*, case study #9–575–060 (Boston: Harvard Business School Publishing, 1985).
6. "USAir Completes Takeover of Pacific Southwest," *Washington Post*, May 30, 1987.
7. The patent-comparison analysis drew from three sources: (1) United States Patent and Trademark Office (USPTO) Official Database, (2) Dialog Research Services, and (3) Professor Jasjit Singh, INSEAD Business School, Patent Citation Analysis Database; "USPTO Patent Full-Text and Image Database," *United States Patent and Trademark Office*, http://www.uspto.gov/; "New Biotechnology Companies," *Science*, February 11, 1983; "Corporate Chronology," *Genentech Inc.*, www.gene.com/gene/about/corporate/history/timeline.html.
8. Noreen Seebacher, "Stryker Products: Just What the Doctor Ordered," *Detroit News*, May 6, 1991, 3F; Barry Stavro, "The Hipbone's Connected to the Bottom Line," *Forbes*, December 3, 1984; Ron Winslow, "As Marketplace Shifts, U.S. Surgical Needs Patching Up," *Wall Street Journal*, February 18, 1994; United States Surgical Corporation, *Fiscal 1987 Annual Report* (Norwalk, CT: United States Surgical Corporation,

1987); Christine Shenot, "U.S. Surgical Innovations Are Cut Above Rest," *Investor's Daily*, March 5, 1991, 36; "Corporate Critics Confidential: Medical Technology," *Wall Street Transcript*, February 11, 1991.

9. "Stakes are Large in Battle for Microprocessor Market," *Globe and Mail*, November 24, 1980, B5; "Section Three: The Great Dark Cloud Falls: IBM's Choice," *CPU Shack*, http://www.cpushack.com/CPU/cpu3.html; George W. Cogan and Robert A. Burgelman, "Intel Corporation (A): The DRAM Decision," *Graduate School of Business, Stanford University*, case study #S–BP–256 (Palo Alto, CA: Graduate School of Business, Stanford University, 1989), 9–10; Ashish Nanda and Christopher A. Bartlett, "Intel Corporation—Leveraging Capabilities for Strategic Renewal," *Harvard Business School*, case study #9–394–141 (Boston: Harvard Business School Publishing, 1994), 3; Aditya P. Mathur, *Introduction to Microprocessor*, 3rd ed. (Noida, India: Tata McGraw-Hill, 1989), 111; "History: 50 Years of Industry Leadership," *National Semiconductor*, www.national.com/analog/company/history; "Intel 8086," *Webster's Online*, www.websters-online-dictionary.org/definitions/Intel+8086?cx=partner-pub-0939450753529744%3Av0qd01-tdlq&cof=FORID%3A9&ie=UTF-8&q=Intel+8086&sa=Search#922; Andrew Pollack, "Intel Offers a 32-Bit Microprocessor," *New York Times*, October 17, 1985; Brenton R. Schlender, "Fast Game: Intel Introduces a Chip Packing Huge Power and Wide Ambitions," *Wall Street Journal*, February 28, 1989.
10. Gerard J. Tellis and Peter N. Golder, *Will and Vision* (New York: McGraw-Hill, 2002), xiii–xv, 43, 46, 290–92.
11. Gene Bylinsky, "How Intel Won Its Bet on Memory Chips," *Fortune*, November 1973, 147, 184, 189; Intel Corporation, *Fiscal 1971 Annual Report* (Santa Clara, CA: Intel Corporation, 1971).
12. Gene Bylinsky, "How Intel Won Its Bet on Memory Chips," *Fortune*, November 1973, 147.
13. Gene Bylinsky, "How Intel Won Its Bet on Memory Chips," *Fortune*, November 1973, 184.
14. "New Leaders in Semiconductors," *Business Week*, March 1, 1976.
15. "New Leaders in Semiconductors," *Business Week*, March 1, 1976.
16. Gene Bylinsky, "How Intel Won Its Bet on Memory Chips," *Fortune*, November 1973, 184; Gordon E. Moore, "Cramming More Components onto Integrated Circuits," *Proceedings of the IEEE*, January 1998 (This is a reprint from the original publication: Gordon E. Moore, "Cramming More Components onto Integrated Circuits," *Electronics*, April 19, 1965.); Leslie Berlin, *The Man Behind the Microchip* (New York: Oxford University Press, 2005), 227; Victor K. McElheny, "High-Technology Jelly Bean Ace," *New York Times*, June 5, 1977; Robert A. Burgelman, Modesto A. Maidique, and Steven C. Wheelwright, *Strategic Management of Technology and Innovation*, 3rd ed. (New York: McGraw-Hill/Irwin, 2001), 931.

17. Personal conversation with author.
18. David Ewing Duncan, *The Amgen Story: 25 Years of Visionary Science and Powerful Medicine* (San Diego: Tehabi Books, 2005), 16, 22–24, 29, 31; James D. Berkley and Nitin Nohria, "Amgen Inc.: Planning the Unplannable," *Harvard Business School*, case study #9–492–052 (Boston: Harvard Business School Publishing, 1992), 2.
19. David Ewing Duncan, *The Amgen Story: 25 Years of Visionary Science and Powerful Medicine* (San Diego: Tehabi Books, 2005), 14, 16, 24, 29, 31, 35, 52–53; Seanna Browder, "How George Rathmann Mastered the Science of the Deal," *Business Week*, November 30, 1998.
20. David Ewing Duncan, *The Amgen Story: 25 Years of Visionary Science and Powerful Medicine* (San Diego: Tehabi Books, 2005), 35.
21. Smith Barney, Harris Upham & Co.; Dean Witter Reynolds Inc.; and Montgomery Securities, "Prospectus: Amgen Common Stock," *Amgen Inc.*, June 17, 1983, 13–17.
22. David Ewing Duncan, *The Amgen Story: 25 Years of Visionary Science and Powerful Medicine* (San Diego: Tehabi Books, 2005), 72, 77–82; Felix Oberholzer-Gee and Dennis Yao, "Amgen Inc.'s Epogen: Commercializing the First Biotech Blockbuster Drug," *Harvard Business School*, case study #7–064–54 (Boston: Harvard Business School Publishing, 2005).
23. Craig E. Aronoff and John L. Ward, *Contemporary Entrepreneurs* (Detroit: Omnigraphics Inc., 1992), 356; Matthew Herper, "Dane Miller: CEO Value to the Bone," *Forbes*, May 8, 2001; Fred R. David, *Strategic Management* (Upper Saddle River, NJ: Prentice Hall, 2003), 376; "Biomet, Inc," *Wall Street Transcript*, January 31, 1994; Richard F. Hubbard and Jeffrey L. Rodengen, *Biomet Inc.: From Warsaw to the World* (Ft. Lauderdale, FL: Write Stuff Enterprises Inc., 2002), 49, 72, 83, 108, 114; "Biomet History," http://www.biomet.com/corporate/biomet Timeline.cfm.
24. Based on year-end 1987 stockholders' equity.
25. M. L. Mead: Scott & Stringfellow Inc., "Kirschner Medical Corporation—Company Report," *The Investext Group*, February 17, 1989, 6; "Kirschner Medical Purchase," *Wall Street Journal*, May 4, 1988; Susan J. Stocker, "After a Dark Year, Kirschner Restores Its New Subsidiary," *Washington Business Journal*, June 19, 1989; Jessica Hall, "Torn Between Two Bidders: Kirschner's Enviable Dilemma," *Warfield's Business Record*, July 1, 1994; L. C. Marsh: Wheat First Butcher & Singer Inc., "Kirschner Medical Corporation—Company Report," *The Investext Group*, October 8, 1990; L. C. Marsh: Wheat First Butcher & Singer Inc., "Kirschner Medical Corporation—Company Report," *The Investext Group*, September 18, 1991; data for the adjacent chart comes from sources included in this note and the previous two end notes.
26. Gary Kissel, *Poor Sailors' Airline* (McLean, VA: Paladwr Press, 2002),

148, 159, 172–73, 186, 193; "Pacific Southwest Airlines," *Wall Street Transcript*, October 20, 1969.

27. "Big Jets Trip Up a Go-Go Airline," *Business Week*, April 14, 1975; Robert Lindsey, "A Fallen Model For Deregulation," *New York Times*, July 13, 1975; PSA Inc., *Fiscal 1970 and 1973 Annual Reports* (San Diego: PSA Inc., 1970 and 1973); Gary Kissel, *Poor Sailors' Airline* (McLean, VA: Paladwr Press, 2002), 173, 179, 193, 196.
28. Gary Kissel, *Poor Sailors' Airline* (McLean, VA: Paladwr Press, 2002), 186–87, 193, 196–97; Robert Lindsey, "A Fallen Model For Deregulation," *New York Times*, July 13, 1975.
29. Richard B. Schmitt and Roy J. Harris Jr., "Braniff-PSA Joint Venture Is Succeeded By Plan to Lease 30 of Grounded Line's Jets," *Wall Street Journal*, n.d.; John S. DeMott, Mark Seal, and Michael Weiss, "Bankruptcy at Braniff," *Time*, May 24, 1982; Gary Kissel, *Poor Sailors' Airline* (McLean, VA: Paladwr Press, 2002), 196, 261, 265, 273–74, 280, 287; Jeffrey M. Lenorovitz, "PSA, Lockheed Sue in L-1011 Dispute," *Aviation Week & Space Technology*, January 8, 1979; Joan M. Feldman, "PSA Switch to DC-9-80 Beginning to Pay Dividends," *Air Transport World*, December 1981; "Death Over San Diego," *Time*, October 9, 1978.
30. Agis Salpukas, "US Air to Buy P.S.A. for $400 Million," *New York Times*, December 9, 1986; Gary Kissel, *Poor Sailors' Airline* (McLean, VA: Paladwr Press, 2002), 301.
31. Katrina Brooker, Herb Kelleher, and Reporter Associate Alynda Wheat, "The Chairman of the Board Looks Back," *Fortune*, May 21, 2001; Tom Krazit, "Intel to Discontinue Rambus Chip Sets," *IDG News*, May 21, 2003; Jeff Chappell, "The Costly Rambus Bandwagon," *Electronic News*, November 6, 2000.
32. The Progressive Corporation, *Fiscal 1986 Annual Report* (Mayfield Village, OH: The Progressive Corporation, 1986), 17, 24; "Like to Drink and Drive?" *Financial World*, November 27, 1990; Nicolaj Siggelkow and Michael E. Porter, "Progressive Corporation," *Harvard Business School*, case study #9–797–109 (Boston: Harvard Business School Publishing, 1998), 15; Gregory E. David, "Chastened?," *Financial World*, January 4, 1994; Jay Greene, "Progressive Corp. High-Risk Insurer Flying High Again," *Plain Dealer*, June 7, 1993.
33. Jay Greene, "Progressive Corp. High-Risk Insurer Flying High Again," *Plain Dealer*, June 7, 1993; Robert G. Knowles, "Progressive Launches Marketing 'Experiment,'" *National Underwriter Property & Casualty-Risk & Benefits Management*, July 22, 1991.
34. Robert G. Knowles, "Progressive Launches Marketing 'Experiment,'" *National Underwriter Property & Casualty-Risk & Benefits Management*, July 22, 1991; Jay Greene, "Progressive Corp. Takes Chance on Standard Coverage," *Plain Dealer*, September 7, 1991; Colleen Mulcahy, "Agents Uneasy with Progressive Auto Contract," *National Underwriter Property & Casualty-Risk & Benefits Management*, September 27, 1993;

James King, "Risk Has Its Rewards," *Plain Dealer*, June 20, 1994, 2S; Frances X. Frei and Hanna Rodriguez-Farrar, "Innovation at Progressive (A): Pay-As-You-Go Insurance," *Harvard Business School*, case study #9–602–175 (Boston: Harvard Business School Publishing, 2004), 4; "Total Auto, Total Premiums Written—2002," *Best's Review*, October 2003; The Progressive Corporation, *Fiscal 1996 Annual Report* (Mayfield Village, OH: The Progressive Corporation, 1996).

35. Frances X. Frei and Hanna Rodriguez-Farrar, "Innovation at Progressive (B): Homeowners Insurance," *Harvard Business School*, case study #9–601–138 (Boston: Harvard Business School Publishing, 2004), 2; Elisabeth Boone, "Recipe for Success," *Rough Notes*, April 2002.
36. "Love is Ammunition for a Texas Airline," *Business Week*, June 26, 1971; Roland Huntford, *The Last Place on Earth* (New York: Modern Library, 1999), 91, 94, 256.
37. James Wallace and Jim Erickson, *Hard Drive* (New York: HarperBusiness, 1992), 172–76.
38. Richard Brandt and Katherine M. Hafner, "The Waiting Game that Microsoft Can't Lose," *Business Week*, September 12, 1988; James Wallace and Jim Erickson, *Hard Drive* (New York: HarperBusiness, 1992), 346–51.
39. James Wallace and Jim Erickson, *Hard Drive* (New York: HarperBusiness, 1992), 349; Richard Brandt and Katherine M. Hafner, "The Waiting Game that Microsoft Can't Lose," *Business Week*, September 12, 1988.
40. Richard Brandt and Katherine M. Hafner, "The Waiting Game that Microsoft Can't Lose," *Business Week*, September 12, 1988; "Gates Reaffirms Faith in OS2," *Computer Weekly*, March 16, 1989; "Windows Keeps Rolling Toward a Career Year," *PC Week*, July 17, 1989.
41. Richard Brandt and Evan I. Schwartz, "IBM and Microsoft: They're Still Talking, But . . ." *Business Week*, October 1, 1990; Philip M. Rosenzweig, "Bill Gates and the Management of Microsoft," *Harvard Business School*, case study #9–392–019 (Boston: Harvard Business School Publishing, 1993); "Microsoft Shipments of Windows Exceed One Million a Month," *Wall Street Journal*, August 12, 1992; Carl Johnston, Michael Rukstad, and David Yoffie, "Microsoft, 2000," *Harvard Business School*, case study #9–700–071 (Boston: Harvard Business School Publishing, 2000), 3; "Microsoft Company," *Operating System*, http://www.operating-system.org/betriebssystem/_english/fa-microsoft.htm; "A History of Windows," *Microsoft Corporation*, http://windows.microsoft.com/en-US/windows/history.
42. Jerry Useem, "Simply Irresistible," *Fortune*, March 19, 2007; "Apple Stores," *ifoAppleStore*, www.ifoapplestore.com/stores/chronology_2001-2003.html.
43. John Markoff, "An 'Unknown' Co-Founder Leaves After 20 Years of Glory and Turmoil," *New York Times*, September 1, 1997; "The Televi-

sion Program Transcripts: Part III—Triumph of the Nerds," *PBS*, www.pbs.org/nerds/part3.html; Gregory C. Rogers and Michael Beer, "Apple Computer (A) (Abridged): Corporate Strategy and Culture," *Harvard Business School*, case study #9–495–044 (Boston: Harvard Business School Publishing, 1997), 4; Mary Kwak and David B. Yoffie, "Apple Computer 1999," Harvard Business School, case study #9–799–108 (Boston: Harvard Business School Publishing, 1999), 6. Source for all stock-return calculations in this work: ©200601 CRSP®, Center for Research in Security Prices. Graduate School of Business, The University of Chicago. Used with permission. All rights reserved. www.crsp.chicagobooth.edu; Jai Singh, "Dell: Apple Should Close Shop," *CNET News*, October 6, 1997.

44. Mary Kwak and David B. Yoffie, "Apple Computer 1999," *Harvard Business School*, case study #9–799–108 (Boston: Harvard Business School Publishing, 1999), 12–13; Gabriel Madway, "Apple CEO-in-Waiting Tim Cook Haunted by Vision Quest," *Reuters*, February 23, 2011; Jim Carlton, *Apple* (New York: Random House, 1997), 15; Apple Inc., *Fiscal 1997 10K* (Cupertino, CA: Apple Inc., 1997); Yusi Wang and David B. Yoffie, "Apple Computer 2002," *Harvard Business School*, case study #9–702–469 (Boston: Harvard Business School Publishing, 2003); Overhead (in this case, selling, general and administrative expenses) decreased as a percentage of sales.
45. Julie Hennessy and Andrei Najjar, "Apple Computer, Inc.: Think Different, Think Online Music," *Kellogg School of Management*, case study #KEL065 (Evanston, IL: Northwestern University, 2004), 2–3, 6; Brent Schlender, "How Big Can Apple Get?" *Fortune*, February 21, 2005; Rob Walker, "The Guts of a New Machine," *New York Times*, November 30, 2003.
46. Julie Hennessy and Andrei Najjar, "Apple Computer, Inc.: Think Different, Think Online Music," *Kellogg School of Management*, case study #KEL065 (Evanston, IL: Northwestern University, 2004), 6; David B. Yoffie and Michael Slind, "Apple Computer, 2006," *Harvard Business School*, case study #9–706–496 (Boston: Harvard Business School Publishing, 2007), 13; Morten T. Hansen, *Collaboration* (Boston: Harvard Business School Publishing, 2009), 7.
47. Apple Inc., *Fiscal 2001–2 10Ks* (Cupertino, CA: Apple Inc., 2001–2).
48. Rob Walker, "The Guts of a New Machine," *New York Times*, November 30, 2003; David B. Yoffie and Michael Slind, "Apple Computer, 2006," *Harvard Business School*, case study # 9–706–496 (Boston: Harvard Business School Publishing, 2007), 17; Julie Hennessy and Andrei Najjar, "Apple Computer, Inc.: Think Different, Think Online Music," *Kellogg School of Management*, case study #KEL065 (Evanston, IL: Northwestern University, 2004), 6–9; Olga Kharif, "iPod: A Seed for Growth," *Business Week*, August 27, 2002.
49. Based on Apple's market share of less than 5 percent in 2001.

50. Yusi Wang and David B. Yoffie, "Apple Computer 2002," *Harvard Business School*, case study #9–702–469 (Boston: Harvard Business School Publishing, 2003); Peter Burrows, "Steve Jobs: 'I'm an Optimist,' " *Business Week Online*, August 13, 2003; "iPod + iTunes Timeline," *Apple Inc.*, www.apple.com/pr/products/ipodhistory/; Julie Hennessy and Andrei Najjar, "Apple Computer, Inc.: Think Different, Think Online Music," *Kellogg School of Management*, case study #KEL065 (Evanston, IL: Northwestern University, 2004), 11; David B. Yoffie and Michael Slind, "Apple Computer, 2006," *Harvard Business School*, case study #9–706–496 (Boston: Harvard Business School Publishing, 2007), 17.
51. Jim Carlton, *Apple: The Inside Story of Intrigue, Egomania, and Business Blunders* (New York: HarperBusiness, 1997), 394–428; Miguel Helft and Ashlee Vance, "Apple Passes Microsoft as No. 1 in Tech," *New York Times*, May 26, 2010.

CHAPTER 5: LEADING ABOVE THE DEATH LINE

1. Quotable Emerson, http://www.quotableemerson.com/allquotes.php.
2. David Breashears, *High Exposure* (New York: Simon & Schuster Paperbacks, 1999), 149, 214, 231, 242, 250–51; "Ed Viesturs on IMAX *Everest*: The Exclusive Mountain Zone Interview," *Mountain Zone*, http://classic.mountainzone.com/climbing/everest/imax/.
3. David Breashears, *High Exposure* (New York: Simon & Schuster Paperbacks, 1999), 149, 250–52.
4. David Breashears, *High Exposure* (New York: Simon & Schuster Paperbacks, 1999), 251–53, 255–56, 261.
5. David Breashears, *High Exposure* (New York: Simon & Schuster Paperbacks, 1999), 224, 232–34, 252–53.
6. David Breashears, *High Exposure* (New York: Simon & Schuster Paperbacks, 1999), 237, 240, 254–55; Jon Krakauer, *Into Thin Air* (New York: Anchor Books, 1997), 68.
7. David Breashears, *High Exposure* (New York: Simon & Schuster Paperbacks, 1999), 284, 289, 291; Jon Krakauer, *Into Thin Air* (New York: Anchor Books, 1997), xv; *National Geographic Adventure*, http://www.nationalgeographic.com/adventure/everest/index.html; "Everest Fatalities," *Adventure Stats*, http://www.adventurestats.com/tables/everest fatilities.shtml.
8. David Breashears, *High Exposure* (New York: Simon & Schuster Paperbacks, 1999), 217, 298; Jon Krakauer, *Into Thin Air* (New York: Anchor Books, 1997), 27, 34–36, 65, 68–69.
9. David Breashears, *High Exposure* (New York: Simon & Schuster Paperbacks, 1999), 224; Jon Krakauer, *Into Thin Air* (New York: Anchor Books, 1997), 153, 207–10; *Storm Over Everest* (Washington, DC: PBS Frontline, 2008), DVD.

10. Jon Krakauer, *Into Thin Air* (New York: Anchor Books, 1997), 171.
11. David Breashears, *High Exposure* (New York: Simon & Schuster Paperbacks, 1999), 217, 232, 261, 265, 281, 284, 289, 296; Jon Krakauer, *Into Thin Air* (New York: Anchor Books, 1997), 208, 214, 258; *Storm Over Everest* (Washington, DC: PBS Frontline, 2008), DVD.
12. Intel Corporation, *Fiscal 1997 and 1999 Annual Reports* (Santa Clara, CA: Intel Corporation, 1997 and 1999); Advanced Micro Devices, *Fiscal 1999 Annual Report* (Sunnyvale, CA: Advanced Micro Devices, 1999). Calculations are based on cash and short-term investments.
13. "Intel: The Microprocessor Champ Gambles on Another Leap Forward," *Business Week*, April 14, 1980, 94; Mimi Real and Robert Warren, *A Revolution in Progress . . . A History of Intel to Date* (Santa Clara, CA: Intel Corporate Communications Department, 1984), 7, 46; Leslie Berlin, *The Man Behind the Microchip* (New York: Oxford University Press, 2005), 172.
14. Steven Rosenbush, Robert D. Hof, and Ben Elgin, "Too Much Cash, Too Little Innovation," *Business Week*, July 18, 2005; Jeremy Quittner, "Entrepreneurs Hoard Cash," *Business Week*, April 16, 2008; Ben McClure, "Cash: Can a Company Have Too Much?" *Investopedia*, http://boards.investopedia.com/articles/fundamental/03/062503.asp.
15. Tim Olper, Lee Pinkowitz, Rene Stulz, and Rohan Williamson, "The Determinants and Implications of Corporate Cash Holdings," *Journal of Financial Economics*, 1999, 17. Note: In financial statements, insurance companies account for liquidity differently from the other industries in our study. For that reason, Progressive's cash-to-asset ratios were excluded from this statistic.
16. Nassim Nicholas Taleb, *The Black Swan* (New York: Random House, 2007); *Nassim N. Taleb Home & Professional Page*, http://www.fooledbyrandomness.com.
17. Southwest Airlines Co., *Fiscal 1991 Annual Report* (Dallas: Southwest Airlines Co., 1991), 3.
18. Terry Maxon, "Southwest Airlines' Chances for Survival Good in Industry Crisis," *Knight Ridder/Tribune Business News*, October 4, 2001; "It Must Be the Peanuts," *CFO*, December 2001, 48; Kim Clark, "Nothing But the Plane Truth," *U.S. News & World Report*, December 31, 2001, 58; "Southwest Airlines Soars with Morningstar's CEO of the Year Award," *PR Newswire*, January 4, 2002; Southwest Airlines Co., *Fiscal 2001 Annual Report* (Dallas: Southwest Airlines Co., 2001), 5; Southwest Airlines Co., *Fiscal 2002 Annual Report* (Dallas: Southwest Airlines Co., 2002), 2.
19. "It Must Be the Peanuts," *CFO*, December 2001, 48; Marc L. Songini, "Southwest Expands Business Tools' Role: Will Manage Operational Data with Tools that Helped Stabilize Finances after Attacks," *Computerworld*, July 15, 2002, 6; Southwest Airlines Co., *Fiscal 2001 Annual Report* (Dallas: Southwest Airlines Co., 2001), 4.

20. *Good to Great*, produced by Sam Tyler (Boston: Northern Light Productions, 2006), DVD.
21. David Breashears, *High Exposure* (New York: Simon & Schuster Paperbacks, 1999), 251–56, 265, 285.
22. Andy Grove with Bethany McClean, "Taking on Prostate Cancer," *Fortune*, May 13, 1996.
23. Data for this paragraph and the adjacent chart: Stryker Corporation, *Fiscal 1989, 1990, 1992, 1994, 1996, 1997, and 1998 Annual Reports* (Kalamazoo, MI: Stryker Corporation, 1989, 1990, 1992, 1994, 1996, 1997, and 1998).
24. Brenda Rios, "Kalamazoo, Mich., Medical Products Firm to Buy Pfizer Orthopedics Unit," *Knight Ridder/Tribune Business News*, August 14, 1998; James P. Miller, "Conservative Stryker Joins the Merger Game in a Big Way," *Wall Street Journal*, August 21, 1998, 1; Stryker Corporation, *Fiscal 1998 Annual Report* (Kalamazoo, MI: Stryker Corporation, 1998), 6.
25. Brenda Rios, "Kalamazoo, Mich., Medical Products Firm to Buy Pfizer Orthopedics Unit," *Knight Ridder/Tribune Business News*, August 14, 1998; James P. Miller, "Conservative Stryker Joins the Merger Game in a Big Way," *Wall Street Journal*, August 21, 1998, 1; Stryker Corporation, *Fiscal 1989, 1996, 1998, and 2000 Annual Reports* (Kalamazoo, MI: Stryker Corporation, 1989, 1996, 1998, and 2000).
26. Personal conversation with author.
27. Daniel J. Simons and Christopher F. Chabris, "Gorillas in Our Midst: Sustained Inattentional Blindness for Dynamic Events," *Perception*, 1999, 1059–70.
28. We found three different spellings of Mr. Buckhout's name. We elected to use the spelling as it appears in Intel's official 15-year history, Mimi Real and Robert Warren, *A Revolution in Progress . . . A History of Intel to Date* (Santa Clara, CA: Intel Corporate Communications Department, 1984), 15.
29. William H. Davidow, *Marketing High Technology* (New York: The Free Press, 1986), 1–11; Mimi Real and Robert Warren, *A Revolution in Progress . . . A History of Intel to Date* (Santa Clara, CA: Intel Corporate Communications Department, 1984), 15.
30. Mimi Real and Robert Warren, *A Revolution in Progress . . . A History of Intel to Date* (Santa Clara, CA: Intel Corporate Communications Department, 1984), 15; William H. Davidow, *Marketing High Technology* (New York: The Free Press, 1986), 4–6.
31. William H. Davidow, *Marketing High Technology* (New York: The Free Press, 1986), 7–8, 10; Katie Woodruff, *Defining Intel: 25 Years/25 Events* (Santa Clara, CA: Intel Corporation, 1993), 16; Tim Jackson, *Inside Intel: Andy Grove and the Rise of the World's Most Powerful Chip Company* (New York: Plume, 1997), 194.
32. Gordon M. Binder, *Amgen* (n.p. The Newcomen Society of the United

States, 1998), 12; David Ewing Duncan, *The Amgen Story: 25 Years of Visionary Science and Powerful Medicine* (San Diego: Tehabi Books, 2005), 84–85.

33. Ellen Benoit, "Breakfast at the Ritz," *Financial World*, March 10, 1987, 18; Marilyn Chase, "FDA Panel Rejection of Anti-Clot Drug Sets Genentech Back Months, Perils Stock," *Wall Street Journal*, June 1, 1987, 26; Jesus Sanchez, "Rejection of Genentech's Heart Drug Surprises Biotechnology Investors," *Los Angeles Times*, June 2, 1987, 1; Stuart Gannes and Gene Bylinsky, "The Big Boys are Joining the Biotech Party: Corporate Giants are about to Crowd the Start-Ups," *Fortune*, July 6, 1987, 58; Andrew Pollack, "Taking the Crucial Next Step at Genentech," *New York Times*, January 28, 1990.
34. Jesus Sanchez, "Rejection of Genentech's Heart Drug Surprises Biotechnology Investors," *Los Angeles Times*, June 2, 1987, 1; Brenton R. Schlender, "Genentech's Missteps and FDA Policy Shift Led to TPA Setback," *Wall Street Journal*, June 16, 1987, 1.
35. Marilyn Chase, "FDA Panel Rejection of Anti-Clot Drug Sets Genentech Back Months, Perils Stock," *Wall Street Journal*, June 1, 1987, 26; Jesus Sanchez, "Rejection of Genentech's Heart Drug Surprises Biotechnology Investors," *Los Angeles Times*, June 2, 1987, 1; "Paradise Postponed," *Economist*, June 6, 1987; "Genentech, Biotechnology Stocks Tumble After Ruling on TPA Drug for Blood Clots," *Wall Street Journal*, June 2, 1987, 3.
36. Reginald Rhein Jr., "FDA Pulls Out the Stops to Approve Genentech's TPA," *Chemical Week*, November 25, 1987, 9.
37. Joan O'C. Hamilton and Reginald Rhein Jr., "A Nasty Shock for Genentech," *Business Week*, June 15, 1987, 37; Reginald Rhein Jr., "FDA Pulls Out the Stops to Approve Genentech's TPA," *Chemical Week*, November 25, 1987, 10.
38. Helen Wheeler, "The Race After Genentech," *High Technology Business*, September 1987, 38, 42; Joan O'C. Hamilton, "Rivals Horn In On Genentech's Heart Drug," *Business Week*, October 26, 1987, 112L.
39. Don Clark, "Genentech May Set Trend: Deal Gives Biotech Firm R&D Money," *San Francisco Chronicle*, February 3, 1990, B1; Jane Fitz Simon, "Swiss Firm to Buy US Biotech Giant," *Boston Globe*, February 3, 1990, 1; Karol Neilsen, "Roche Floats Genentech Shares," *Chemical Week*, November 17, 1999, 33; Andrew Pollack, "Roche Offers $43.7 Billion for Shares in Genentech It Does Not Already Own," *New York Times*, July 22, 2008, 6; "Swiss Drug Giant Roche Buys Up Genentech," *CBS News*, March 12, 2009, http://www.cbsnews.com/stories/2009/03/12/business/main4861008.shtml?source=RSSattr=Health_4861008. Source for all stock-return calculations in this work: ©200601 CRSP®, Center for Research in Security Prices. Booth School of Business, The University of Chicago. Used with permission. All rights reserved. www.crsp.chicagobooth.edu.

40. Roland Huntford, *The Last Place on Earth* (New York: Random House, 1999), 197, 202, 204–6; Roald Amundsen, *The South Pole* (McLean, VA: IndyPublish.com, 2009), 25–26.
41. Roland Huntford, *The Last Place on Earth* (New York: Random House, 1999), 284–85, 288; Roald Amundsen, *The South Pole* (McLean, VA: IndyPublish.com, 2009), 70–72, 205–7, 346.

CHAPTER 6: SMaC

1. *Le Malade Imaginaire*, Act III scene iii.
2. Howard D. Putnam with Gene Busnar, *The Winds of Turbulence* (Reno, NV: Howard D. Putnam Enterprises Inc., 1991), 8, 12–14, 302.
3. Howard D. Putnam, "Southwest Airlines Co.: Presentation by Howard D. Putnam, President and Chief Executive Officer, Before the Dallas Association of Investment Analysts," *Wall Street Transcript*, May 28, 1979; "Texas Gets Bigger," *Forbes*, November 12, 1979.
4. "Icelandair," *Funding Universe*, "http://www.fundinguniverse.com/company-histories/Icelandair-Company-History.html.
5. "Fact Sheet: Fleet," *Southwest Airlines Co.*, http://www.southwest.com/html/about-southwest/history/fact-sheet.html.
6. James Wallace and Jim Erickson, *Hard Drive* (New York: HarperBusiness, 1992), 54, 491–92; Noreen Seebacher, "Stryker Products: Just What the Doctor Ordered," *Detroit News*, May 6, 1991, 3F; Michael Tubbs, "Recession is a Chance to Increase R&D Expenditure," *Financial Times*, December 2, 2008; Barry Stavro, "Amgen Plays It Cool Despite Clamor Over EPO," *Los Angeles Times*, June 7, 1989; James Ellis, "You Don't Necessarily Get What You Pay For," *Business Week*, May 4, 1992.
7. David Breashears, *High Exposure* (New York: Simon & Schuster, 1999), 217–18, 225, 294–96; David Breashears, "David Breashears Speech Preview," *YouTube*, http://video.google.com/videoplay?docid=5383977496159243481#; Personal conversation with author.
8. David Breashears, *High Exposure* (New York: Simon & Schuster, 1999), 23, 217, 219, 224, 232, 239, 245, 249, 265, 281, 285, 295; David Breashears, "David Breashears Speech Preview," *YouTube*, http://video.google.com/videoplay?docid=5383977496159243481#.
9. Robert G. Knowles, "Progressive Launches Marketing 'Experiment,'" *National Underwriter Property & Casualty-Risk & Benefits Management*, July 22, 1991; "Thomas A. King: The Progressive Corporation (PGR)," *Wall Street Transcript*, January 14, 2002; Peter Lewis, "The Progressive Corporation: Address by Peter B. Lewis, President to the New York Society of Security Analysis," *Wall Street Transcript*, January 24, 1972; Amy Hutton and James Weber, "Progressive Insurance: Disclosure Strategy," *Harvard Business School*, case study #9–102–012 (Boston: Harvard Business School Publishing, 2002), 3–4; Gregory

David, "Chastened?" *Financial World,* January 4, 1994, 40; Frances X. Frei and Hanna Rodriguez-Farrar, "Innovation at Progressive (A): Pay-As-You-Go Insurance," *Harvard Business School,* case study #9–602–175 (Boston: Harvard Business School Publishing, 2002), 5; The Progressive Corporation, *Fiscal 1986 Annual Report* (Mayfield Heights, OH: The Progressive Corporation, 1986), 17–18; The Progressive Corporation, *Fiscal 2001 Annual Report* (Mayfield Village, OH: The Progressive Corporation, 2001), 20; Nicolaj Siggelkow and Michael E. Porter, "Progressive Corporation," *Harvard Business School,* case study #9–797–109 (Boston: Harvard Business School Publishing, 1998), 8; The Progressive Corporation, *Fiscal 1971 Annual Report* (Cleveland, OH: The Progressive Corporation, 1971); Robert McGough, "Like to Drink and Drive?" *Financial World,* November 27, 1990, 27.

10. Gary Kissel, *Poor Sailors' Airline* (McLean, VA: Paladwr Press, 2002), 221, 231, 295; Jerry Brown, "PSA President: Sale of AirCal Sparked Merger," *Travel Weekly,* December 18, 1986.
11. Southwest Airlines Co., *Fiscal 1987 Annual Report* (Dallas: Southwest Airlines Co., 1987); James E. Ellis, "These Two Airlines Are Doing It Their Way," *Business Week,* September 21, 1987; "Southwest Airlines Company (LUV)," *Wall Street Transcript,* June 13, 1988.
12. Julie Pitta, "Apple's Mr. Pragmatist," *Forbes,* March 28, 1994.
13. Jim Carlton, *Apple: The Inside Story of Intrigue, Egomania, and Business Blunders* (New York: Random House, 1997), 13–14, 20–21; Michael Gartenberg, "Now Apple's Really 'For the Rest of Us,' " *Macworld.com,* June 23, 2010; Mary Kwak and David B. Yoffie, "Apple Computer 1999," *Harvard Business School,* case study #9–799–108 (Boston: Harvard Business School Publishing, 1999), 2–5; Johanna M. Hurstak and David B. Yoffie, "Reshaping Apple Computer's Destiny 1992," *Harvard Business School,* case study #9–393–011 (Boston: Harvard Business School Publishing, 1992), 5; Julie Pitta, "Apple's Mr. Pragmatist," *Forbes,* March 28, 1994; John Markoff, "An 'Unknown' Co-Founder Leaves After 20 Years of Glory and Turmoil," *New York Times,* September 1, 1997; Chris Preimesberger, "How Apple Dodged a Sun Buyout," *eWeek,* http://www.eweek.com/c/a/IT-Infrastructure/How-Apple-Dodged-a-Sun-Buyout-Former-CEOs-McNealy-Zander-Tell-All-251679/. Source for all stock-return calculations in this work: ©200601 CRSP®, Center for Research in Security Prices. Booth School of Business, The University of Chicago. Used with permission. All rights reserved. www.crsp.chicagobooth.edu. Data for the chart "1985–1997: Microsoft Soared, Apple Faltered" included in this end note.
14. Brent Schlender, "How Big Can Apple Get?" *Fortune,* February 21, 2005.
15. Brent Schlender, "How Big Can Apple Get?" *Fortune,* February 21, 2005; James Pomfret and Kelvin Soh, "For Apple Suppliers, Loose Lips Can Sink Contracts," *Reuters,* February 17, 2010; Devin Leonard,

"Songs in the Key of Steve," *Fortune*, May 12, 2003; Julie Hennessy and Andrei Najjar, "Apple Computer, Inc.: Think Different, Think Online Music," *Kellogg School of Management*, case study #KEL065 (Evanston, IL: Northwestern University, 2004), 16; Nick Wingfield, "Core Value: At Apple, Secrecy Complicates Life But Maintains Buzz," *Wall Street Journal*, June 28, 2006; David Kirkpatrick and Tyler Maroney, "The Second Coming of Apple," *Fortune*, November 9, 1998; Thomas E. Weber, "Why I Fired Steve Jobs," *Daily Beast*, June 6, 2010.

16. Rick Bernstein and Ross Greenburg, *The UCLA Dynasty* (New York: Home Box Office Inc, 2008), DVD.
17. Rick Bernstein and Ross Greenburg, *The UCLA Dynasty* (New York: Home Box Office Inc, 2008), DVD.
18. "Abraham Lincoln," *Quotations Book*, http://quotationsbook.com/quote/44576/#axzz1JL6NjMqm.
19. Kathleen K. Wiegner, "Why A Chip Is Not A Chip," *Forbes*, June 17, 1985; Mary Bellis, "Inventors of the Modern Computer: Intel 4004—The World's First Single Chip Microprocessor," *About.com*, http://inventors.about.com/od/mstartinventions/a/microprocessor.htm; Dan Steere and Robert A. Burgelman, "Intel Corporation (D): Microprocessors at the Crossroads," *Graduate School of Business, Stanford University*, case study #BP–256D (Palo Alto, CA: Graduate School of Business, Stanford University, 1994).
20. Bruce Graham and Robert A. Burgelman, "Intel Corporation (B): Implementing the DRAM Decision," *Graduate School of Business, Stanford University*, case study #S–BP–256B (Palo Alto, CA: Graduate School of Business, Stanford University, 1991), 1.
21. Gordon E. Moore, "Cramming More Components onto Integrated Circuits," *Proceedings of the IEEE*, January 1998 (This is a reprint from the original publication: Gordon E. Moore, "Cramming More Components onto Integrated Circuits," *Electronics*, April 19, 1965.); "Intel: Supplier Rising as a Big Competitor," *New York Times*, February 14, 1990, D1; Robert N. Noyce, "Large-Scale Integration: What is Yet to Come?" *Science*, March 1977; Ramon Casadesus-Masanell and David B. Yoffie, "Intel Corporation: 1968–2003 (Teaching Note)," *Harvard Business School*, case study #5–704–465 (Boston: Harvard Business School Publishing, 2004), 2; "Craig Barrett is Leading the Chip Giant Into Riskier Terrain," *Business Week*, March 13, 2000, 110; Leslie Berlin, *The Man Behind the Microchip* (New York: Oxford University Press, 2005), 227; Gene Bylinsky, "How Intel Won Its Bet on Memory Chips," *Fortune*, November 1973, 184; Don Clark, "Intel Lawyer Commands Chip War," *San Francisco Chronicle*, June 28, 1993; Andrew S. Grove, "How to Make Confrontation Work For You," *Fortune*, July 13, 1984; "Creativity by the Numbers: An Interview with Robert N. Noyce," *Harvard Business Review*, May–June 1980; "IBM and Intel Link Up to Fend Off Japan,"

Business Week, January 10, 1983; Tim Jackson, *Inside Intel* (New York: Penguin Putnam Inc., 1997), 9, 313–16; Don Clark, "Inside Intel, It's All Copying," *Wall Street Journal*, October 28, 2002.

22. Jeffrey L. Rodengen, *The Spirit of AMD: Advanced Micro Devices* (Fort Lauderdale, FL: Write Stuff Enterprises Inc., 1998), 55, 67–68, 90–92; Advanced Micro Devices, *Fiscal 1998 and 2002 Annual Reports* (Sunnyvale, CA: Advanced Micro Devices, 1998 and 2002).
23. Mary Bellis, "Inventors of the Modern Computer: Intel 4004—The World's First Single Chip Microprocessor," *About.com*, http://inventors.about.com/od/mstartinventions/a/microprocessor.htm.
24. J. Allard, "Windows: The Next Killer Application on the Internet," interoffice memo, *Microsoft*, January 25, 1994, www.microsoft.com/about/ . . . /docs/di_killerapp_InternetMemo.rtf; Kathy Rebello, Amy Cortese, and Rob Hof, "Inside Microsoft: The Untold Story of How the Internet Forced Bill Gates to Reverse Course," *Business Week*, July 15, 1996, 35–40; Bill Gates, "How I Work: Bill Gates," *Fortune*, April 7, 2006.
25. Bill Gates, "The Internet Tidal Wave," May 26, 1995, http://www.justice.gov/atr/cases/exhibits/20.pdf; Kathy Rebello, Amy Cortese, and Rob Hof, "Inside Microsoft: The Untold Story of How the Internet Forced Bill Gates to Reverse Course," *Business Week*, July 15, 1996, 38; Brent Schlender and Sheree R. Curry, "Software Hardball Microsoft is Spending Billions to Crush Netscape and Control the Internet," *Fortune*, September 30, 1996.
26. Lester B. Orfield, "Federal Amending Power: Genesis and Justiciability," *Minnesota Law Review*, 1930, 369–84; "The United States Constitution: Amendments," *U.S. Constitution Online*, http://www.usconstitution.net/; "Bill of Rights and Later Amendments," *Independence Hall Association*, www.ushistory.org/.
27. "Centuries of Citizenship: A Constitutional Timeline," *National Constitution Center*, http://constitutioncenter.org/timeline/html/cw02.html; Declaration of Independence, *USHistory.org*, http://www.ushistory.org/declaration/document/.

CHAPTER 7: RETURN ON LUCK

1. Marshall Bruce Mathers III (Eminem), "Lose Yourself," *8 Mile* (soundtrack), Universal Import, 2002, compact disc.
2. Sources for the Malcolm Daly story told throughout this chapter are as follows: Malcolm Daly, "Malcolm Daly's Accident on Thunder Mountain in the Alaska Range on 5/19/99," *Trango*, http://www.trango.com/stories/mal_accident.pdf; Dave Krupa, "Jim Donini (Interview)," *Denali National Park Jukebox Series*, June 30, 2000; personal conversations with author, February 2010–April 2011; "Non-Profit Helps Disabled Enjoy Outdoors," *Sierra Blogging Post*, http://blog.sierra

tradingpost.com/in-outdoors-camping-gear-forest-trails/non-profit-helps-disabled-enjoy-the-outdoors/.

3. "Donini Bags Three Patagonian First Ascents," *The American Alpine Club*, January 12, 2009, http://www.americanalpineclub.org/news/doninibagsthree; "Jim Donini," *Wikipedia*, http://en.wikipedia.org/wiki/Jim_Donini; "Jack Tackle: Professional Biography," *Dirty Sox Club*, http://dirtysoxclub.wordpress.com/members/jack-tackle/.
4. Gordon M. Binder, *Amgen* (n.p.: The Newcomen Society of the United States, 1998), 10.
5. James Bates, "Biotech Detective Scores Coup: Amgen Scientist Spent Years Searching for the Key to Producing EPO," *Los Angeles Times*, June 2, 1989, 1.
6. Peter Behr, "Boom or Bust in the Biotech Industry," *Environment*, July/August 1982, 6; Steve Curwood, "Biotech Bellyache," *Boston Globe*, August 23, 1983, 1.
7. James Bates, "Biotech Detective Scores Coup: Amgen Scientist Spent Years Searching for the Key to Producing EPO," *Los Angeles Times*, June 2, 1989, 1; Gordon M. Binder, *Amgen* (n.p.: The Newcomen Society of the United States, 1998), 13.
8. Alun Anderson and David Swinbanks, "Growing Pains for Amgen as Epoetin Wins US Approval," *Nature*, June 1989, 493; Edmund L. Andrews, "Mad Scientists," *Business Month*, May 1990, 54; Edmund L. Andrews, "Patents; Unaddressed Question in Amgen Case," *New York Times*, March 9, 1991.
9. Henry Gee, "Amgen Scores a Knockout," *Nature*, March 1991, 99; Barry Stavro, "Court Upholds Amgen's Patent on Anemia Drug Medicine," *Los Angeles Times*, March 7, 1991, 1; Edmund L. Andrews, "Amgen Wins Fight Over Drug," *New York Times*, March 7, 1991, D1; Rhonda L. Rundle and David Stipp, "Amgen Wins Biotech Drug Patent Battle," *Wall Street Journal*, March 7, 1991, A3; Elizabeth S. Kiesche, "Amgen Wins EPO Battle, but Patent War Goes On," *Chemical Week*, March 20, 1991, 16; Paul Hemp, "High Court Refuses Genetics Patent Appeal," *Boston Globe*, October 8, 1991, 39.
10. Wade Roush, " 'Fat Hormone' Poses Hefty Problem for Journal Embargo," *Science*, August 4, 1995, 627; Larry Armstrong, John Carey, and Geoffrey Smith, "Will This Drug End Obesity?" *Business Week*, August 7, 1995, 29; Christiane Truelove, "Bio Biotech, Big Pharma," *Med Ad News*, September 1999, 50; David Ewing Duncan, *The Amgen Story: 25 Years of Visionary Science and Powerful Medicine* (San Diego: Tehabi Books, 2005), 135–36, 157.
11. David Ewing Duncan, *The Amgen Story: 25 Years of Visionary Science and Powerful Medicine* (San Diego: Tehabi Books, 2005), 135–36, 157.
12. Paul A. Gompers, "The Rise and Fall of Venture Capital," *Business and Economic History*, Winter 1994, 2; Carl T. Hall, "Biotechnology Revolution—20 Years Later," *San Francisco Chronicle*, May 28, 1996, B1.

13. "Investors Dream of Genes," *Time*, October 20, 1980, 72.
14. Ron Scherer, "Wall Street's Wild Fling with Hot High Tech," *Christian Science Monitor*, October 17, 1980, 1, 17; Robert Lenzner, "Taking Stock," *Boston Globe*, October 19, 1980, 1; Adam Lashinsky, "Remembering Netscape: The Birth of the Web," *Fortune*, July 25, 2005; Douglas MacMillan, "Google's Historic IPO: Beatable," *Business Week*, August 16, 2007; "Corporate Chronology," *Genentech Inc.*, http://www.gene.com/gene/about/corporate/history/timeline.html.
15. "Robert A. Swanson '70 (1947–1999)," *MIT Entrepreneurship Center—Legendary Leaders and Memorials*, http://entrepreneurship.mit.edu/legendary_leaders_memorials.php; "Timeline of Biotechnology," *Biotechnology Institute*, http://www.biotechinstitute.org/what-is-biotechnology/timeline?tid=103.
16. Ray Snoddy, "Genentech Push to Manufacturing," *Financial Times*, June 18, 1982, 13; William D. Marbach, Pamela Abramson, Robb A. Allan, Cynthia Rigg, and Phyllis Malamud, "The Bust in Biotechnology," *Newsweek*, July 26, 1982, 73; Peter Behr, "Boom or Bust in the Biotech Industry," *Environment*, July/August 1982, 6; Genentech Inc., *Fiscal 1985 Annual Report* (San Francisco: Genentech Inc., 1985).
17. Jerry E. Bishop, "Genentech Seeks Human Tests of Drug to Dissolve Clots During Heart Attacks," *Wall Street Journal*, November 16, 1983, 60; "Licensing of Activase Marks New Era in Treating Heart Attacks," *Genentech, Inc. Press Releases*, November 13, 1987, http://www.gene.com/gene/news/press-releases/display.do?method=detail&id=4271; Marilyn Chase, "Genentech Expected to Post Strong Net for 1987, Spurred by Launch of TPA," *Wall Street Journal*, January 12, 1988, 1.
18. Nell Henderson, "Biotech Breakthrough Focuses on Heart Attacks; Survival Tech Works on Delivering New Drug," *Washington Post*, October 12, 1986, H1.
19. Andrea Gabor and Peter Dworkin, "Superdrugs from Genetic Secrets," *U.S. News & World Report*, March 24, 1986, 54; Joan O'C. Hamilton, "Biotech's First Superstar," *Business Week*, April 14, 1986, 68; Louise Kehoe, "Fresh Blood and New Heart; Eagle Eye," *Financial Times*, January 19, 1988, 32.
20. The TIMI Study Group, "Comparison of Invasive and Conservative Strategies after Treatment with Intravenous Tissue Plasminogen Activator in Acute Myocardial Infarction," *New England Journal of Medicine*, March 9, 1989, 320, 618–27; Lawrence K. Altman, "Study Finds No Difference in 2 Heart Attach [sic] Drugs," *New York Times*, March 30, 1989; Michael Waldholz, "Genentech Heart Drug Dealt Critical Blow—Head to Head Study Finds TPA Is Only As Effective As Rival Streptokinase," *Wall Street Journal*, March 30, 1989.
21. Michael Waldholz, "Heart Attack Study May Spur Use of Less-Costly TPA Alternative," *Wall Street Journal*, August 12, 1988, 1; Richard L. Hudson, "Genentech's Heart Drug TPA Appears Only to Equal Its Ri-

vals, Report Says," *Wall Street Journal*, September 2, 1988, 1; Marilyn Chase, "Lost Euphoria: Genentech, Battered by Great Expectations, Is Tightening Its Belt," *Wall Street Journal*, October 11, 1988, 1; Marilyn Chase, "Little Difference is Found Between TPA and Rival in Small Study of Heart Drugs," *Wall Street Journal*, November 14, 1988, 1; Michael Waldholz, "Genentech Heart Drug Dealt Critical Blow—Head to Head Study Finds TPA Is Only As Effective As Rival Streptokinase," *Wall Street Journal*, March 30, 1989, 1; "Genentech's Fortunes: A Boost for CD4 and 'Crosscurrents' in 1988," *Pharmaceutical Business News*, April 14, 1989; Sabin Russell, "Heart-Attack Drug Study is a Blow to Genentech," *San Francisco Chronicle*, March 9, 1990, A1; "Heart-Attack Drugs: Trials and Tribulations," *Economist*, March 16, 1991, 86; Genentech Inc., *Fiscal 1989 Annual Report* (San Francisco: Genentech Inc., 1989).

22. Charles Petit, "Genentech Beats Cheaper Rival in Battle of Heart Attack Drugs," *San Francisco Chronicle*, May 1, 1993, A1; Genentech Inc., *Fiscal 1994 Annual Report* (San Francisco: Genentech Inc., 1994).
23. Gene Bylinsky, "How Intel Won Its Bet on Memory Chips," *Fortune*, November 1973, 184; Thomas C. Hayes, "Intel's Earnings Grew Sharply in Fourth Quarter," *New York Times*, January 14, 1984.
24. Gordon M. Binder, *Amgen* (n.p.: The Newcomen Society of the United States, 1998), 10.
25. James Bates, "Biotech Detective Scores Coup: Amgen Scientist Spent Years Searching for the Key to Producing EPO," *Los Angeles Times*, June 2, 1989, 1; David Ewing Duncan, *The Amgen Story: 25 Years of Visionary Science and Powerful Medicine* (San Diego: Tehabi Books, 2005), 66, 71; Edmund L. Andrews, "Mad Scientists," *Business Month*, May 1990, 54; Pamela Sherrid, "Biotech Battle Royale," *U.S. News & World Report*, March 20, 2000, 52; "Billion Dollar Babies: Biotech Drugs as Blockbusters," *Nature Biotechnology*, April 2007.
26. White, Weld & Co. and McDonald & Company, "Prospectus: The Progressive Corporation Common Stock," *The Progressive Corporation*, April 15, 1971; April Dougal Gasbarre (updated by David Bianco), "The Progressive Corporation," *International Directory of Company Histories* (New York: St. James Press, 1999), 397; The Progressive Corporation, *Fiscal 2000 Annual Report* (Mayfield Village, OH: The Progressive Corporation, 2000).
27. James Wallace and Jim Erickson, *Hard Drive* (New York: HarperBusiness, 1992), 20, 27, 53, 67, 71–76.
28. According to Table 1, page 23, in the Bukoski and Korotkin article cited here, in 1970, 3.9 percent of U.S. public secondary schools used computers for instruction. William J. Bukoski and Arthur L. Korotkin, "Computing Activities in Secondary Education," *American Institutes for Research in the Behavioral Sciences*, September 1975, 2–30; Andrew

Molnar, "Computers in Education: A Brief History," *THE Journal*, June 1, 1997.

29. Andrew Molnar, "Computers in Education: A Brief History," *THE Journal*, June 1, 1997.
30. James Wallace and Jim Erickson, *Hard Drive* (New York: HarperBusiness, 1992), 76–77, 97, 110.
31. Jeffrey L. Rodengen, *The Spirit of AMD: Advanced Micro Devices* (Ft. Lauderdale, FL: Write Stuff Enterprises Inc., 1998), 127; Stephen Kreider Yoder, "Changing Game: Intel Faces Challenge to Its Dominance in Microprocessors," *Wall Street Journal*, April 8, 1991, A1; Ken Siegmann, "Intel Loses Copyright Suit Against Rival," *San Francisco Chronicle*, March 11, 1994, A1; Jim Carlton and Stephen Kreider Yoder, "Computers: Humble Pie: Intel to Replace its Pentium Chips," *Wall Street Journal*, December 21, 1994, B1; Don Clark, "Intel's 4th Period Net Fell 37% on Big Charge for Pentium Woes," *Wall Street Journal*, January 18, 1995, B6; Stewart Alsop and Patty de Llosa, "Can AMD Avoid the Intel Graveyard?" *Fortune*, April 14, 1997; Ira Sager and Andy Reinhardt, "Chipping at Intel's Lead," *Business Week*, October 19, 1998, 46; Advanced Micro Devices, *Fiscal 1994 and 1998 Annual Reports* (Sunnyvale, VA: Advanced Micro Devices, 1994 and 1998.
32. Advanced Micro Devices, *Fiscal 1995 Annual Report* (Sunnyvale, VA: Advanced Micro Devices, 1995); Jeffrey L. Rodengen, *The Spirit of AMD: Advanced Micro Devices* (Ft. Lauderdale, FL: Write Stuff Enterprises Inc., 1998), 133–36.
33. Jim Carlton, "Advanced Micro Woos a Partner to Fight Intel," *Wall Street Journal*, October 23, 1995, A3; Robert D. Hof and Peter Burrows, "Intel Won't Feel the Heat from this Fusion," *Business Week*, November 6, 1995; "My Chip is Faster than Your Chip," *Business Week*, February 10, 1997, 70; "Advanced Micro Lands Deal with Digital," *Dow Jones Online News*, April 25, 1997; Jeffrey L. Rodengen, *The Spirit of AMD: Advanced Micro Devices* (Ft. Lauderdale, FL: Write Stuff Enterprises Inc., 1998), 137–39; Ira Sager and Andy Reinhardt, "Chipping at Intel's Lead," *Business Week*, October 19, 1998, 46; "Semiconductors: The Monkey and the Gorilla," *Economist*, December 5, 1998, 71.
34. Dean Takahashi, "More Bad News Puts Intel Rival Further Behind," *Wall Street Journal*, June 24, 1999, B1; Angela Key, "Hello (Again), Mr. Chips," *Fortune*, April 3, 2000; "Semiconductors: The Monkey and the Gorilla," *Economist*, December 5, 1998, 71; Jeffrey L. Rodengen, *The Spirit of AMD: Advanced Micro Devices* (Ft. Lauderdale, FL: Write Stuff Enterprises Inc., 1998), 133–36, 141. Source for all stock-return calculations in this work: ©200601 CRSP®, Center for Research in Security Prices. Booth School of Business, The University of Chicago. Used with permission. All rights reserved. http://www.crsp.chicagobooth.edu.
35. James Wallace and Jim Erickson, *Hard Drive* (New York: HarperBusiness, 1992), 167, 173, 175–77, 179–81; Lisa Miller Mesdag, "Famous

Victories in Personal Software," *Fortune*, May 2, 1983, 153; Julia Pitta, "Coulda Been a Contender," *Forbes*, July 10, 1989; John Markoff, "PC Software Maker Novell to Buy Digital Research," *New York Times*, July 17, 1991.

36. James Wallace and Jim Erickson, *Hard Drive* (New York: HarperBusiness, 1992) 176, 190; Lisa Miller Mesdag, "Famous Victories in Personal Software," *Fortune*, May 2, 1983, 153; Julia Pitta, "Coulda Been a Contender," *Forbes*, July 10, 1989.
37. Marcia Stepanek, "Q&A with Progressive's Peter Lewis," *Business Week*, September 12, 2000; "About Us: Provisions of Proposition 103 Affecting the Rate Regulation Division," *California Department of Insurance*, http://www.insurance.ca.gov/0500-about-us/0500-organization/0400-rate-regulation/prop-103.cfm; The Progressive Corporation, *Fiscal 1991 Annual Report* (Mayfield Heights, OH: The Progressive Corporation, 1991).
38. Stephen Phillips, "Driven to Succeed Peter Lewis, Progressive's Artful Chief Exec, Aims to Overtake Auto Insurance Industry's Leaders," *Plain Dealer*, September 1, 1996, 1.I; "Ralph Nader Biography," *Academy of Achievement*, http://www.achievement.org/autodoc/page/nad0bio-1; James Wallace and Jim Erickson, *Hard Drive* (New York: HarperBusiness, 1992), 76; Brian Dumaine, "Times are Good? Create a Crisis," *Fortune*, June 28, 1993, 123; David Craig, "Progressively Thinking," *USA Today*, September 15, 1994, 01.B; Carol J. Loomis, "Sex. Reefer? And Auto Insurance!" *Fortune*, August 7, 1995, 76.
39. Betsy Wiesendanger, "Progressive's Success is No Accident," *Sales & Marketing Management*, September 1991, 57; Ronald Henkoff, "Service is Everybody's Business," *Fortune*, June 27, 1994, 48; Carol J. Loomis, "Sex. Reefer? And Auto Insurance!" *Fortune*, August 7, 1995, 76; "Leading Writers of Private Passenger Auto Insurance," *Best's Review*, September 1988, 22; "All Private Passenger Auto," *Best's Review*, October 1998; "Gearing Up: Insurers are Using Driver Safety Programs, Sharply Focused Advertising and the Internet to Court Teen Drivers," *Best's Review*, October 2003; Chuck Salter, "Progressive Makes Big Claims," *Fast Company*, November 1998, 176.
40. Malcolm Gladwell, *Outliers*, paperback edition (New York: Back Bay Books / Little Brown and Company, 2011), 20–30.
41. Our calculations are included in the *Research Foundations* section. It is worth noting that there is some discrepancy and ambiguity in the data cited in *Outliers* by author Malcolm Gladwell. In the paperback edition of *Outliers* (New York: Back Bay Books / Little Brown and Company, 2011), Gladwell wrote on pages 22–23 about statistics gathered by Canadian psychologist Roger Barnsley and A. H. Thompson: "He [Barnsley] looked at the composition of the National Hockey League. Same story. The more he looked, the more Barnsley came to believe that what he was seeing was not a chance occurrence but an iron law of

Canadian hockey: in *any* elite group of hockey players—the very best of the best—40 percent of the players will have been born between January and March, 30 percent between April and June, 20 percent between July and September, and 10 percent between October and December." In the endnotes to the paperback edition of *Outliers*, Gladwell provided the following citation: "Roger Barnsley and A. H. Thompson have put their study on a Web site, http://www.socialproblemindex.ualberta.ca/relage.htm." That website page referenced an original article by Barnsley and Thompson as follows: "Source: Barnsley RH, Thompson AH, Barnsley PE (1985). Hockey success and birth-date: The relative age effect. *Journal of the Canadian Association for Health, Physical Education, and Recreation*, Nov.–Dec., 23–28." We tracked down a paper copy of that original source article in library archives (we were unable to find it online); the NHL data table in that article ("Table 2, Months of Birth, National Hockey League Players, 1982/83 Season" on page 24) gives the distribution as 32.0% / 29.8% / 21.9% / 16.2%. Even so, our counterargument to the birth-date argument would still hold: for the truly elite Hall of Fame players, any relative advantage that accrues to youth-league players born in the first part of the year melts away. The greatest players find a way to make themselves into 10Xers regardless of birth date.

42. "Athlete Profile: Ray Bourque," *Sports Illustrated*, February 3, 1998, http://sportsillustrated.cnn.com/olympics/events/1998/nagano/athletes/235.htm; "Ray Bourque," *National Hockey League*, http://www.nhl.com/ice/player.htm?id=8445621; "Ray Bourque," *HockeyDB.com*, http://www.hockeydb.com/ihdb/stats/pdisplay.php?pid=520; Robin Finn, "Bourque: A Star Without the Sparkle," *New York Times*, February 3, 1986; Joe Lapointe, "Hockey: Bourque, at 33, is Still Mr. Defense," *New York Times*, January 21, 1994; "Athlete Profile: Ray Bourque," *Sports Illustrated*, February 3, 1998, http://sportsillustrated.cnn.com/olympics/events/1998/nagano/athletes/235.htm; Nancy Marrapese-Burrell, "The Clock Chimes for Father Ice Time," *ESPN*, http://espn.go.com/classic/biography/s/Bourque_Ray.html; "One on One with Ray Bourque," *Hockey Hall of Fame*, http://www.legendsofhockey.net; *NHL Stats*, http://www.nhl.com/ice/statshome.htm.
43. *Greensboro Youth Hockey Association*, http://www.gyhastars.com/Page.asp?n=9340&org=gyhastars.com.
44. Friedrich Nietzsche, *Twilight of the Idols* (Indianapolis: Hackett Publishing Company, 1997). Note: This book was originally published in 1888.
45. Lamar Muse, *Southwest Passage: The Inside Story of Southwest Airlines' Formative Years* (Austin, TX: Eakin Press, 2002), 92.
46. "Pacific Southwest Airlines: Speech by Paul C. Barkley, President, to the Society of Airline Analysts of the New York Society of Security Analysts, June 23, 1982," *Wall Street Transcript*, August 9, 1982.
47. Gary Kissel, *Poor Sailors' Airline* (McLean, VA: Paladwr Press, 2002),

234, 245, 262, 281, 283, 291, 295; Howard D. Putnam with Gene Busnar, *The Winds of Turbulence* (Reno, NV: Howard D. Putnam Enterprises Inc., 1991), 206–7; "PSA Inc. Debt Rating is Lowered by Moody's," *Wall Street Journal*, October 7, 1982, 41; "PSA Plans Layoffs, Melding of Operations," *Wall Street Journal*, December 1, 1983; "PSA's Airline Warns of Closing if Workers Make No Concessions," *Wall Street Journal*, November 12, 1984, 1.

48. Richard F. Hubbard and Jeffrey L. Rodengen, *Biomet Inc: From Warsaw to the World* (Ft. Lauderdale, FL: Write Stuff Enterprises Inc., 2002), 12–29; David Cassak, "Biomet's Contrarian Conservatism," *Business and Medicine Report*, May 1999.

EPILOGUE: GREAT BY CHOICE

1. F. Scott Fitzgerald, *The Crack-Up* (New York: New Directions; 1945), 57.

FREQUENTLY ASKED QUESTIONS

1. David Breashears, *High Exposure* (New York: Simon & Schuster Paperbacks, 1999), 285; Sally B. Donnelly, "One Airline's Magic: How Does Southwest Soar Above its Money-Losing Rivals? Its Employees Work Harder and Smarter, in Return for Job Security and a Share of the Profits," *Time*, October 28, 2002, 45; Robert McGough, "Like to Drink and Drive?" *Financial World*, November 27, 1990, 27; The Progressive Corporation, *Fiscal 1981 Annual Report* (Mayfield Village, OH: The Progressive Corporation, 1981), 11; Noreen Seebacher, "Stryker Products: Just What the Doctor Ordered," *Detroit News*, May 6, 1991, 3F; Geoffrey Brewer, "20 Percent—Or Else!" *Sales & Marketing Management*, November 1994; Barry Stavro, "Amgen Bets Its Future on Biotech Anemia Drug," *Los Angeles Times*, May 12, 1987, 9A; Gordon M. Binder, "Amgen," *The Newcomen Society of the United States*, 1998, 19; Tom Wolfe, "The Tinkerings of Robert Noyce," *Esquire Magazine*, December 1983, 346–74; Leslie Berlin, *The Man Behind the Microchip* (New York: Oxford University Press Inc., 2005), 151, 157, 163; James Wallace and Jim Erickson, *Hard Drive* (New York: HarperBusiness, 1992), 260; Rich Karlgaard, "ASAP Interview: Bill Gates (Microsoft Corp.'s CEO)," *Forbes*, December 7, 1992; Julia Lawlor, "Microsoft's Rite of Spring," *USA Today*, April 8, 1993, 01B; Geoffrey Smith and James Ellis, "Pay That was Justified—And Pay that Just Mystified," *Business Week*, May 6, 1991.
2. William Patrick Patterson, "Software Sparks a Gold Rush," *Industry Week*, October 17, 1983; Dennis Kneale, "Overload System: As Software Products and Firms Proliferate, A Shakeout is Forecast," *Wall Street Journal*, February 23, 1984; "25-Year PC Anniversary Statistics," *Com-*

puter Industry Almanac Inc., http://www.c-i-a.com/pr0806.htm; Michael Miller, "More Than 1 Billion Sold," *PCMag.com*, August 6, 2002, http://www.pcmag.com/article2/0,2817,427042,00.asp.

RESEARCH FOUNDATIONS

1. Organizational behavior research has drawn on the following insights about the case-research methodology: Juliet M. Corbin and Anselm C. Strauss, *Basics of Qualitative Research: Techniques and Procedures for Developing Grounded Theory* (Thousand Oaks, CA: Sage Publications, 2008, 3rd ed). Robert K. Yin, *Case Study Research: Design and Methods* (Thousand Oaks, CA: Sage Publications, 2009, 4th ed). Matthew B. Miles and A. Michael Huberman, *Qualitative Data Analysis: An Expanded Sourcebook* (Thousand Oaks, CA: Sage Publications, 1994, 2nd ed).

   As an example from finance research, Mark Chen, in his article in *Journal of Finance* in 2004, analyzed why some companies adopted restrictions on repricing of options, while others did not. He first picked companies that had restrictions on repricing options (selecting on the dependent or outcome variable) and then matched each of those firms with another (comparison) based on similarities in industry, firm size, and year of observation. (Mark A. Chen, "Executive Option Repricing, Incentives, and Retention," *Journal of Finance*, June 2004, 1167–99.) Other finance studies using matched samples include Bélen Villalonga, "Does Diversification Cause the 'Diversification Discount'?," *Financial Management*, Summer 2004, 5–27; and Kenneth Lehn and Annette Poulsen, "Free Cash Flow and Stockholder Gains in Going Private Transactions," *Journal of Finance*, July 1989, 771–87.

   Examples of studies in medicine using the pairwise-control sample method include Andrew D. Shaw et al., "The Effect of Aprotinin on Outcome after Coronary-Artery Bypass Grafting," *New England Journal of Medicine*, February 2008, 784–93; and Jack V. Tu et al., "Effectiveness and Safety of Drug-Eluting Stents in Ontario," *New England Journal of Medicine*, October 2007, 357:1393–1402.
2. See Kathleen M. Eisenhardt and Melissa E. Graebner, "Theory Building from Cases: Opportunities and Challenges," *Academy of Management Journal*, February 2007, 25–32.
3. Jeffrey A. Martin and Kathleen M. Eisenhardt, "Rewiring: Cross-Business-Unit Collaborations in Multibusiness Organizations," *Academy of Management Journal*, April 2010, 265–301. From a set of collaborative projects taking place in six firms, they selected two projects in each company, one with high and one with low performance (selecting on the dependent or outcome variable). The pair of projects in each firm was matched on similarities in size, resources, duration, complexity, importance, and type. Based on these matched pairs, they conducted a

qualitative-contrast analysis to generate new insights about what might explain this difference in outcomes.

4. In academic research, selecting on the dependent or outcome variable is sometimes seen as poor research design; however, that assessment is based on a confusing use of terminology. What is really meant by that is *selecting on success*, which means selecting on *one value only* (success) of the dependent variable and ignoring other cases, including cases involving failures. But that is very different from selecting on *different values* (success *and* non-success) of the dependent variable—which is our approach. We agree that studying successful companies only is a limitation and that is why we include comparison companies (average or poor performers) in our study.
5. The two sources were Gary Kissel, *Poor Sailors' Airline: A History of Pacific Southwest Airlines* (McLean, VA: Paladwr Press, 2002), 171–72, and Lamar Muse, *Southwest Passage: The Inside Story of Southwest Airlines' Formative Years* (Austin, TX: Eakin Press, 2002), 84.
6. The classical text in organization sociology on event-history analysis is Nancy B. Tuma and Michael T. Hannan, *Social Dynamics: Models and Methods* (Orlando, FL: Academic Press, 1984).
7. This point is well made in the Eisenhardt and Graebner article (2007) referenced earlier.
8. Robert K. Yin, *Case Study Research: Design and Methods* (Newbury Park, CA: Sage Publications, 2008, 4th ed).
9. There is a long line of research in social psychology regarding how people make incorrect attributions about what causes events, including self-serving attributions, whereby individuals tend to attribute causes of positive outcomes to their own actions and causes of bad outcomes to external factors. A very good overview can be found in Lee Ross and Richard E. Nisbett, *The Person and the Situation: Perspectives of Social Psychology* (New York: McGraw-Hill, 1991).
10. This was a primary finding in Jim Collins, *How the Mighty Fall: And Why Some Companies Never Give In* (Boulder, CO: Jim Collins, 2009).
11. We elected to look at companies limited by the CRSP database for the following reasons:

    - CRSP tracks all publicly traded securities on the NYSE, AMEX, and NASDAQ. We believe that this represents a significant universe of companies from which to select companies that went public during the time period of our study.
    - CRSP is the most reliable source of consistent stock return data, so that we can do rigorous apples-to-apples performance comparisons. To assess whether a company meets the financial criteria for selection into our study *requires* that it appears in the CRSP database.

We decided not to use an alternative database, SDC, because we discovered several weaknesses with it with respect to our needs.

12. There were 3,001 companies that all came online in the CRSP database on the same date: December 29, 1972. This huge swell of data entries on a single date represents the addition of NASDAQ listed companies to the CRSP database in one fell swoop. We wondered if perhaps our 3X the market requirement would unfairly penalize some companies that went public months or a couple of years before the 12/29/72 date. Since the 12/29/72 date is an arbitrary accident of history, we wondered if these companies might be arbitrarily excluded by this onetime anomaly of the data. To address this issue, we tracked companies that came on line on 12/29/72 and whose ratio to the market across all months was between 2X and 3X (46 companies total). When we carefully examined these 46, we found no strong performers that would make it through the other cuts into our study. They were eliminated for any number of reasons: lack of revenues, erratic patterns, flat or declining return ratio patterns, or inability to determine an IPO date. Additionally, we felt that any company that could not somehow maintain at least a 3X ratio of returns over the period 1972 to 1992 would be a questionable candidate, regardless of its return pattern prior to 12/29/72. We decided, therefore, to make no special allowance for the companies that came online on 12/29/72, holding them to the same 3X standard as all the other companies in the final elimination process.
13. "Match year" was the same year as IPO for Microsoft (1986), Amgen (1983), and Stryker (1979). Intel, Southwest Airlines, and Progressive had their IPOs in 1971, but they came online in the CRSP database only in January 1973, which is their match year. For Biomet, we used the match year in which the comparison (Kirschner) came online in CRSP (1986) because this was later than the year when Biomet came online (1983).
14. More precisely, the period of analysis runs from the first month for which there was available stock-return data for the 10X company in the CRSP database to June 28, 2002. The first month was often the month after the 10X company went public (or later in the cases of Southwest, Progressive, Intel, all of which were included in CRSP only from January 1973 onward). We used Kirchner's IPO date (1986), as it was after Biomet's.
15. There are a number of academic studies that have drawn distinctions between degrees and types of innovation. Some classic studies include Rebecca M. Henderson and Kim B. Clark, "Architectural Innovation: The Reconfiguration of Existing Product Technologies and the Failure of Established Firms," *Administrative Science Quarterly*, March 1990, 9–30; Michael L. Tushman and Philip Anderson, "Technological Discontinuities and Organizational Environments," *Administrative Science Quarterly*, September 1986, 439–65; Clayton M. Christensen, *The In-*

novator's Dilemma: When New Technologies Cause Great Firms to Fail (Boston: Harvard Business School Press, 1997).

16. These also include incremental innovations, although very small innovations and tweaks to existing products would likely not hit this radar screen (thus we don't know whether there was a difference between companies in terms of number of very incremental innovations).
17. Source: United States Patent and Trademark Office (USPTO) official database. (http://patft.uspto.gov/netahtml/PTO/search-adv.htm). We counted the number of patents *issued* by year (using the number of patents *applied for* by year leads to the same conclusion). Note that this count includes all patents issued from year of founding (Amgen in 1980, Genentech in 1976) to 2002, while the chart showing patent data in Chapter 7 covers 1983 to 2002.
18. In addition to using patent count to measure a firm's innovativeness (a measure that is very sensitive to a company's proclivity to file patents), management scholars often examine the *average number of citations these patents get from subsequent patents.* This helps better understand the relative quality or significance of each of these patents. We asked Professor Jasjit Singh to pull data on patent citations for Amgen and Genentech from his extensive database on patent citations for thousands of firms worldwide. Although a variety of patent-citation measures exist, we used a simple one. The information he provided us included the number of times a patent had been cited and showed the mean citation count by patent. While a few patents were cited a lot, most received a low number of citations, with an average patent citation of 6.6 in Singh's database. Singh wrote to us, "Even within their specific technology class, Genentech is above average—the normalized citation count is 1.1 for it (1.0 implies you are at average). Amgen again below average at 0.78" (personal communication, June 2, 2010). See the following article for a similar application of patent citations and description of Singh's database: Jasjit Singh, "Distributed R&D, Cross-Regional Knowledge Integration and Quality of Innovative Output," *Research Policy*, 2008, 77–96.
19. Because bullets are by definition small-scale experiments, they may not always show up in written company or public documents. Thus it's quite possible that we underreported the existence of bullets; however, this potential bias would be the same for all the companies in our data set. This wasn't a problem in coding cannonballs, which we found easier to spot as they were big, visible investments and efforts.
20. A number of scholars have addressed the role of luck in society, and in financial markets in particular. These works inspired us to consider luck as an explicit variable in our analysis. While many works exist, Nassim Nicholas Taleb's two books are some of the most well known in this genre: *Fooled by Randomness: The Hidden Role of Chance in Life and in the Markets* (New York: Random House, 2005, 2nd ed.) and *The Black*

Swan: The Impact of the Highly Improbable (New York: Random House, 2010, 2nd ed.).

21. The Hockey Hall of Fame, http://www.hhof.com.
22. Specifically, we divided the group of Canadian-born inductees who were born *before* 1950 and who had played at least one season in the NHL into the following birth groups: 1873–99 (*N* = 21), 1900–29 (*N* = 63), and 1930–49 (*N* = 36). The January–March births were in the same proportion in the Canadian-born-inductee group as they were in the general Canadian population, except for the 1900–1929 cohort, for which 15.9 percent of the inductees were born in January (versus about 8 percent for the general population), although 12.7 percent of the inductees were born in December (versus about 8 percent for the general population). For the 1930–49 cohort, the months with the most inductees were January (13.9 percent), August (13.9 percent), and December (13.9 percent), indicating no bias toward the January–March period. In short, even when we looked back further in time, it was difficult to find a clear trend showing a disproportionate number of inductees born in January or from January to March.
23. Frank Trovato and Dave Odynak provided us with raw Canadian census data from 1926 to 1981, which was one of the data sets used in their journal article, Frank Trovato and Dave Odynak, "The Seasonality of Births in Canada and the Provinces 1881–1989: Theory and Analysis," *Canadian Studies in Population*, 1993, 1–41. We used unadjusted figures. These and other numbers used in other studies are very similar. See, for example, the original study of the relative age effect among hockey players: Roger H. Barnsley, Angus H. Thompson, and Paula E. Barnsley, "Hockey Success and Birth-Date: The Relative Age Effect," *Journal of the Canadian Association for Health, Physical Education, and Recreation*, November 1985, 23–28.

INDEX

Note: Page numbers in *italics* refer to charts.